Lecture Notes in Computer Science 843

Edited by G. Goos and J. Hartmanis

Advisory Board: W. Brauer   D. Gries   J. Stoer

T0216776

Andrzej Szepietowski

Turing Machines with Sublogarithmic Space

Springer-Verlag

Berlin Heidelberg New York
London Paris Tokyo
Hong Kong Barcelona
Budapest

Series Editors

Gerhard Goos
Universität Karlsruhe
Postfach 69 80
Vincenz-Priessnitz-Straße 1
D-76131 Karlsruhe, Germany

Juris Hartmanis
Cornell University
Department of Computer Science
4130 Upson Hall
Ithaca, NY 14853, USA

Author

Andrzej Szepietowski
Mathematical Institute, Gdansk University
Ul. Wita Stwosza 57, 80-952 Gdansk, Poland

Until March 1995:
FB 17 – Informatik, Universität GH Paderborn
Warburgerstraße 100, D-33095 Paderborn, Germany

CR Subject Classification (1991): F.1.1, F.4.1

ISBN 3-540-58355-6 Springer-Verlag Berlin Heidelberg New York
ISBN 0-387-58355-6 Springer-Verlag New York Berlin Heidelberg

CIP data applied for

© Springer-Verlag Berlin Heidelberg 1994
Printed in Germany

Typesetting: Camera-ready by author
SPIN: 10475435   45/3140-543210 - Printed on acid-free paper

Preface

The main objects of investigation of space complexity theory are Turing machines with bounded space (the number of cells on the tape used during computation) and languages accepted by such machines. Although some important problems still remain open, much work has been done, and many exciting results have been obtained. Many of these results, however, have been proved under the assumption that the amount of available space is at least logarithmic with respect to the length of the input.

In this book I have presented the key result on space complexity, but concentrated on problems: What happens when we drop the assumption of at least logarithmic space and what do languages acceptable with sublogarithmic space look like?

The manuscript for this book was written at the Technical University of Gdansk and Gdansk University from 1991 to 1992. Parts of the manuscript were used as notes for lectures given at these universities. I want to thank the participants in these courses for their criticism. The final corrections to the manuscript were made during my stay at the University of Paderborn in 1994 under an Alexander von Humboldt Research Fellowship.

I would like to express my profound gratitude to Professor A.W. Mostowski for his encouragement and advice.

I owe a special debt to an anonymous referee for his or her helpful comments, to Mrs. K. Mostowska and J. Skurczyński for their careful reading of the original version of the manuscript, and to L. Chańko for helping me with typesetting the manuscript in TEX.

Paderborn, May 1994   Andrzej Szepietowski

Contents

1 **Introduction** 1

2 **Basic Notions** 7
   2.1 Turing Machines 7
   2.2 Configuration and Computation 9
   2.3 Internal Configuration 11
   2.4 Alternating Turing Machines 12
   2.5 Space Complexity 13

3 **Languages Acceptable with Logarithmic Space** 15
   3.1 Two Examples of Languages Acceptable with
       Logarithmic Space 15
   3.2 Pebble Automata 16
   3.3 $NSPACE(\log n)$ Complete Language 18

4 **Examples of Languages Acceptable with Sublogarithmic
Space** 21
   4.1 Languages Acceptable with $o(\log n)$ Space 21
   4.2 Tally Languages Acceptable with Sublogarithmic Space 24

5 **Lower Bounds for Accepting Non-regular Languages** 27
   5.1 Lower Bounds for Two-way Turing Machines 28
   5.2 Lower Bounds for One-way Turing Machines 33
   5.3 Lower Bounds for the Middle Mode of Space Complexity 34

6 **Space Constructible Functions** 37
   6.1 Definitions and Basic Properties 37
   6.2 Fully Space Constructible Functions 39
   6.3 Nondeterministically Fully Space Constructible Functions .. 42

7 **Halting Property and Closure under Complement** 47
   7.1 Halting Property of Turing Machines with Logarithmic or
       Greater Space 48
   7.2 Closure under Complement of Strong Deterministic
       Complexity Classes 49
   7.3 Closure under Complement of Strong Nondeterministic Space
       Complexity Classes above Logarithm 52
   7.4 Closure under Complement for Bounded Languages Acceptable
       by Nondeterministic Turing Machines 53

7.5 Nonclosure under Complement of Language Acceptable by
Weakly Space Bounded Turing Machines. 55

8 Strong versus Weak Mode of Space Complexity 61
8.1 Weak and Strong Mode of Space Complexity for Fully Space
Constructible Functions 61
8.2 Weak and Strong Complexity Classes above Logarithm 62
8.3 Weak and Strong Complexity Classes below Logarithm 63
8.4 Strong Mode of Space Complexity for Recognizing Machines . 66
8.5 Weak and Middle Modes of Space Complexity above Logarithm 66

9 Padding 67
9.1 Padding above Logarithm 67
9.2 Padding below Logarithm 70

10 Deterministic versus Nondeterministic Turing Machines .. 77
10.1 Determinism versus Nondeterminism above Logarithm 77
10.2 Determinism versus Nondeterminism below Logarithm 78

11 Space Hierarchy 81
11.1 Diagonalization 81
11.2 Space Hierarchy below Logarithm 82

12 Closure under Concatenation 85

13 Alternating Hierarchy 89
13.1 Collapsing of the Alternating Hierarchy above Logarithm ... 90
13.2 Noncollapsing of the Alternating Hierarchy below Logarithm 92

14 Independent Complement 95

15 Other Models of Turing Machines 99
15.1 Two-dimensional Turing Machines 100
15.2 Inkdot Turing Machines 104
15.3 1-pebble Turing Machines 107
15.4 Demon Turing Machines 109

References 111

Subject Index 113

Symbol Index 115

1. Introduction

This monograph is an investigation of the computational power of Turing machines with sublogarithmic space. We consider the Turing machine model introduced by Stearns, Hartmanis, and Lewis (1965) with a two-way, read-only input tape and a separate two-way, read-write work tape. The number of tape cells used on the work tape, called space, is our measure of computational complexity. Two definitions of space complexity have been used in literature. Let $L(n)$ be a function on the natural numbers. A Turing machine is said to be *strongly $L(n)$* space-bounded if no computation on any input of length n, uses more than $L(n)$ space. It is said to be *weakly $L(n)$* space-bounded if for every accepted input of length n, at least one accepting computation (or tree for alternating Turing machines) uses no more than $L(n)$ space. $DSPACE[L(n)]$, $NSPACE[L(n)]$, or $ASPACE[L(n)]$ denotes the class of languages accepted by deterministic, nondeterministic, or alternating $L(n)$ space-bounded Turing machines, respectively. These classes are called space complexity classes, deterministic, nondeterministic, or alternating, respectively.

A great number of problems concerning space complexity still remain open. We do not know much about the relationship between deterministic and nondeterministic space complexity classes. But recently, there have been some very exciting developments in the study of space-bounded computations. In this book we review the key results about space complexity but concentrate on the specific properties and proof techniques required in the study of languages accepted within sublogarithmic space, i.e. space bounded by a function $L(n)$ such that $L(n) = o(\log n)$ or $\liminf \frac{L(n)}{\log n} = 0$.

Turing machines with sublogarithmic space differ sharply from those which have logarithmic or greater space. In this book we shall discuss many examples of this fact and to illustrate our point we present below some of them briefly.

1. Classes of languages accepted within sublogarithmic space depend heavily on the machine models and the mode of space complexity. Many models of Turing machine with bounded space are equivalent to each other if the space bound function is equal to or greater than logarithm. Hence, the results concerning logarithmic or greater space are, in some sense, model independent. This is not the case for sublogarithmic space. However, there are also model independent results in the case of sublogarithmic space. For example, the function $\log \log n$ is the lower bound for accepting non-regular languages for many different models of Turing machine and modes of space complexity.

2. Many results that hold for logarithmic or greater space are not valid for sublogarithmic space. For example: (i) The alternating hierarchy for $SPACE[L(n)]$

collapses to Σ_1 level if $L(n) \geq \log n$, and is infinite if $L(n) = o(\log n)$ and $L(n) \geq \log \log n$. (ii) If $L(n) \geq \log n$, then the class of languages accepted within $L(n)$ space is closed under concatenation. This is not the case if $L(n) = o(\log n)$. (iii) Above $\log n$, all "normal" functions are fully space constructible. Below $\log n$, no unbounded non-decreasing function is fully space constructible.

3. Many results that hold for both logarithmic and sublogarithmic space complexity classes require more sophisticated proof techniques in the case of sublogarithmic space. Compare, for example, the elegant proof of (Sipser 1980) that the deterministic $L(n)$ space complexity class is closed under complement for an arbitrary function $L(n)$, with the analogous proof for $L(n) \geq \log n$. Also the new padding argument used in (Szepietowski 1990) showing that if the deterministic and nondeterministic space complexity classes are equal for the space bound $\log \log n$, then they are also equal for $\log n$, is more complicated than the standard padding argument of (Savitch 1970).

Now we shall outline what is to come in the subsequent chapters.

Chapter 2 contains basic concepts and definitions. In Chapter 3 we give some examples of languages acceptable with logarithmic space and we show that Turing machines with logarithmic space are equivalent to pebble automata. We also present the language GAP (graph accessibility problem) and prove that GAP is complete for the class $NSPACE(\log n)$. Chapter 4 gives some examples of non-regular languages accepted within sublogarithmic space.

Chapter 5 presents so-called gap theorems. In one of the first papers on space complexity Stearns et. al. (1965) proved that it is impossible for a Turing machine to lay off small amounts of tape; either it lays off an amount bounded by a constant or the amount laid off must reach $d \log \log n$ infinitely often for some constant d. Hence, $\log \log n$ is a lower bound of space complexity for accepting non-regular languages. Afterwards $\log \log n$ was proved to be the lower bound for other models of Turing machines: deterministic, nondeterministic, and alternating, and both kinds of space complexity: weak and strong, (Hopcroft and Ullman 1967; Alberts 1985; Iwama 1993).

We shall also present results of (Stearns et. al. 1965) and (Alberts 1985) concerning lower bounds for accepting non-regular languages by one-way Turing machines, i.e. the Turing machines whose input head can only move to the right. The lower bound for one-way Turing machines is equal to $\log n$ for strongly space-bounded deterministic, nondeterministic, and alternating, and weakly space-bounded deterministic Turing machines, and is equal to $\log \log n$ for weakly space-bounded nondeterministic and alternating Turing machines.

Chapter 6 considers space constructible, fully space constructible, and non-deterministically fully space constructible functions. A function $L(n)$ is space constructible if there exists an $L(n)$ space-bounded deterministic Turing machine M such that for every n, there exists an input of length n on which M uses exactly $L(n)$ space. $L(n)$ is fully space constructible if M uses exactly $L(n)$ space on every input of length n. Finally a function $L(n)$ is nondeterministically fully space constructible if there exists a strongly $L(n)$ space-bounded nondeterministic Turing machine M' such that on every input of length n, there is a computation of M' which stops and uses exactly $L(n)$ space.

Above $\log n$ all "normal" functions are fully space constructible, e.g. $\log n$ itself, n^r, $[\log n]^q$, and q^n with rational $r > 0$ and $q \geq 1$. Nevertheless, we shall show in Chapter 8 that there are computable functions above logarithm which are not fully space constructible. Below $\log n$, the situation is more complicated. Freedman and Ladner (1975) proved that neither $\log\log n$ nor any other unbounded non-decreasing $o(\log n)$ function is fully space constructible, and Geffert (1991) proved the same for nondeterministically fully space constructible functions. The crucial lemma of (Geffert 1991) says that if we consider a nondeterministic Turing machine with a one-letter input alphabet, then we only need to consider those computations which loop when they traverse through the whole input. This lemma has many useful consequences.

In Chapter 7 we consider the halting property of Turing machines and closure under complement of space complexity classes. Using the well-know technique of (Immerman 1988; Szelepcsényi 1988) and of (Sipser 1980) we shall show that every strongly $L(n)$ space-bounded Turing machine (deterministic with arbitrary $L(n)$ and nondeterministic with $L(n) \geq \log n$) can be simulated by a strongly $L(n)$ space-bounded recognizing Turing machine (deterministic or nondeterministic, respectively). A recognizing Turing machine has two special states s_a and s_r and reaches the state s_a on every accepted input, and the state s_r on every not accepted input, and on no input both s_a and s_r can be reached.

Using this fact we shall show that classes $strong\text{-}DSPACE[L(n)]$ for arbitrary $L(n)$ and $strong\text{-}NSPACE[L(n)]$ for $L(n) \geq \log n$, are closed under complement (Sipser 1980; Immerman 1988; Szelepcsényi 1988).

It is not known, whether the class $strong\text{-}NSPACE[L(n)]$ is closed under complement, if $L(n) = o(\log n)$, and whether any strongly $L(n)$ space-bounded nondeterministic Turing machine M can be simulated by a recognizing Turing machine. But due to the result of (Geffert 1991) such simulation by recognizing Turing machines is possible for arbitrary function $L(n)$, if the language acceptable by M is bounded, i.e., if $L(M) \subseteq a_1^* a_2^* \ldots a_k^*$. Therefore, if a language A of the form $A \subseteq a_1^* a_2^* \ldots a_k^*$ is accepted by a strongly $L(n)$ space-bounded nondeterministic Turing machine, then its complement A^c is also accepted by a strongly $L(n)$ space-bounded nondeterministic Turing machine. This means, in particular, that the class of tally languages accepted by strongly $L(n)$ space-bounded nondeterministic Turing machines is closed under complement.

On the other hand, weak complexity classes are generally not closed under complement. The class $weak\text{-}DSPACE[L(n)]$ for arbitrary $L(n)$, and $weak\text{-}NSPACE[L(n)]$ for $L(n) \geq \log n$, are closed under complement only if they are equal to the strong complexity classes. Furthermore, there exist deterministic and nondeterministic weak space complexity classes above logarithm which are not closed under complement; and all weak space complexity classes between $\log\log n$ and $\log n$ are not closed under complement.

Chapter 8 discusses the relations between the weak and strong modes of space complexity. These modes are equivalent if the space bound function is fully space constructible. Thus, they are equivalent for all "normal" functions above logarithm. But there exist weak deterministic and nondeterministic space complexity classes, even above logarithm, which are not equal to the strong ones. We shall also show that between $\log \log n$ and $\log n$, deterministic, nondeterministic and alternating weak space complexity classes are not equal to the strong ones.

The question whether the $L(n)$ space-bounded nondeterministic Turing machines are more powerful than the deterministic ones is a major open question in the space complexity theory. The answer to this question is known only for some special functions $L(n)$ and models of computation. As we have already mentioned, if $L(n) = o(\log \log n)$, then deterministic, nondeterministic, and even alternating Turing machines accept only regular languages. One can also construct, using diagonalization, a total recursive function $L(n)$ such that the deterministic and nondeterministic $L(n)$ space complexity classes are not equal (Hopcroft and Ullman 1979).

In Chapter 9 we show that the question whether the nondeterministic and deterministic space complexity classes are equal for "large" function can be reduced to the analogous question for "small" function. We use the padding argument of (Savitch 1970) to prove that for any "well-behaved" function $L(n) \geq \log$, if the deterministic and nondeterministic space complexity classes are equal for space bound $L(n)$, then they are also equal for any space-bound function $H(n) \geq L(n)$. The padding argument of (Savitch 1970) does not work below logarithm and in (Szepietowski 1990) a new padding argument was used to show that if the deterministic and nondeterministic space complexity classes are equal for $\log \log n$, then they are also equal for $\log n$.

Chapter 10 considers the cost of determinization of nondeterministic Turing machines with bounded space. More precisely, we shall try to find out how much space is needed for the deterministic Turing machine to accept the language which is accepted by a nondeterministic Turing machine in $L(n)$ space. We present the famous result of (Savitch 1970), who proved that if $L(n) \geq \log n$, then every $L(n)$ space-bounded nondeterministic Turing machine can be simulated by an $[L(n)]^2$ space-bounded deterministic Turing machine. Monien and Sudborough (1982) proved that if $L(n) \leq \log n$ and $L(n)$ is fully space constructible or $L(n) = f(\log n)$ for some fully space constructible $f(n)$, then every $L(n)$ space-bounded nondeterministic Turing machine can be simulated by an $L(n) \cdot \log n$ space-bounded deterministic Turing machine. In particular, every $\log \log n$ space-bounded nondeterministic Turing machine can be simulated by a $\log n \cdot \log \log n$ space-bounded deterministic Turing machine. From the result of (Geffert 1991) it follows that for an arbitrary $L(n)$, if a language $A \subseteq a_1^* a_2^* \ldots a_k^*$ is accepted by a strongly $L(n)$ space-bounded nondeterministic Turing machine, then it can also be accepted by a strongly $[L(n)]^2$ space-bounded deterministic Turing machine.

Chapter 11 shows that the space hierarchy is dense, or in other words, that more languages can be accepted with more space. We use diagonalization to prove that for every fully space constructible function $L(n)$, there exists a language $A \subset \{1\}^*$ which is accepted by some $L(n)$ space-bounded deterministic (or nondeterministic) Turing machine M_s and not accepted by any strongly $L_1(n)$ space-bounded deterministic (or nondeterministic, respectively) Turing machine if $L_1(n) = o(L_2(n))$. Chapter 11 also contains the proof of Hopcroft and Ullman (1967) that deterministic and nondeterministic space hierarchy is dense below logarithm.

Chapter 12 discusses closure under concatenation of the space complexity classes. It is easy to see that if $L(n)$ is monotone and $L(n) \geq \log n$, then the $L(n)$ space complexity class is closed under concatenation. This is not the case for sublogarithmic space. Ibarra and Ravikumar (1988) presented a proof that deterministic sublogarithmic space complexity classes are not closed under concatenation. We show that this is also true for nondeterministic sublogarithmic space complexity classes.

In Chapter 13 we show that the alternating hierarchy for $SPACE[L(n)]$ collapses to Σ_1 level if $L(n) \geq \log n$. This fact follows from the theorem of (Immerman 1988; Szelepcsényi 1988) that if $L(n) \geq \log n$, then the nondeterministic strong $L(n)$ space complexity class is closed under complement. In contrast to this, the alternating hierarchy does not collapse to Σ_1 level if $L(n) = o(\log n)$. Taking these two facts into consideration, one can raise a natural question: does this mean that the nondeterministic sublogarithmic space complexity classes are not closed under complement? We discuss this problem in Chapter 14 and although we do not provide a complete answer, we prove some weaker result, viz. that if $L(n) = o(\log n)$, then the class of strongly $L(n)$ space-bounded nondeterministic Turing machines (even with a one-letter input alphabet) is not closed under so-called independent (of the starting configuration) complement. We say that an $L(n)$ space-bounded Turing machine M_c is an independent (of the starting configuration) complement of another $L(n)$ space-bounded Turing machine M if for each input x and each $L(|x|)$ space-bounded configuration β of M, M_c starting on x in β accepts if and only if M starting on x in β does not accept. We also show that there exists an $L(n)$ space-bounded nondeterministic Turing machine (even with a one-letter input alphabet) which cannot be simulated in a similar strong way by any $L(n)$ space-bounded deterministic Turing machine (Szepietowski 1989).

Recently von Braunmühr, Gengler, and Rettinger (1994) and Liśkiewicz and Reischuk (1994) proved that the alternating hierarchy for $SPACE[L(n)]$ is infinite if $L(n) \geq \log \log n$ and $L(n) = o(\log n)$.

In Chapter 15 we consider some other models of Turing machines with bounded space and see which functions are space constructible by these machines. We consider two-dimensional Turing machines, 1-inkdot Turing machines, 1-pebble Turing machines, and demon Turing machines. An input tape of a two-dimensional Turing machine is a two-dimensional rectangular array of symbols from some alphabet Σ. A 1-inkdot Turing machine is a usual Turing machine with the additional power of marking one cell on the input tape (with an inkdot). This cell is marked once and for all (no erasing) and no more than one dot of ink is available. A 1-pebble Turing machine is equipped with a pebble which can be placed on and removed from any cell of the input tape. A demon $L(n)$ Turing machine is a Turing machine that has $L(n)$ space marked off on its work tape for every input of length n, and cannot use more space than $L(n)$.

These models do not differ much from the normal Turing machine if the space bound function $L(n) \geq \log n$, but they behave differently if $L(n) = o(\log n)$. This shows that Turing machines with sublogarithmic space are sensitive to modifications of the definition.

In (Szepietowski 1992) it was shown that the function $\log \log n$ is fully space constructible, and functions: $\log^{(k)} n$ for every k, and $\log^* n = \min\{i \mid \log^{(i)} n \leq 1\}$ are space constructible by two-dimensional Turing machines. Thus, the space hierarchy of two-dimensional languages starts much lower than in the one-dimensional case. Hartmanis and Ranjan (1989) showed that the function $\log \log n$ can be fully space constructed by a nondeterministic 1-inkdot Turing machine and that for 1-inkdot Turing machines, nondeterministic sublogarithmic space is more powerful than deterministic one.

2. Basic Notions

In this chapter we present basic concepts and definitions. For more details, we refer to any of the standard books on foundations of theoretical computer science (Hopcroft and Ullman 1979; Balcázar et. al. 1988).

An *alphabet* is any non-empty, finite set. We shall use Σ to denote an alphabet. A *symbol*, or also a *letter*, is an element of an alphabet. A *word* over Σ is a finite sequence of symbols from Σ. The *empty word* λ is the unique word consisting of zero symbols. The *length* of a word w, denoted $|w|$, is the number of symbols w consists of; $w(i)$ denotes the i-th symbol of w. The set of all words over an alphabet Σ will be denoted by Σ^*. A *language* over Σ is a subset of Σ^*. $A \cup B$, $A \cap B$, and $A - B$ denote the *union*, *intersection*, and *difference*, respectively, of two languages A and B. The *complement* of a language A over Σ is the language $A^c = \Sigma^* - A$. $A \subseteq B$ denotes the *inclusion*. By $A \subsetneq B$ we shall denote the *proper inclusion*, i.e. $A \subseteq B$ but $A \neq B$. The *power set* of A, denoted $\mathcal{P}(A)$, is the set of all subsets of A. The *concatenation* of A and B, denoted AB, is the set $\{xy \mid x \in A \text{ and } y \in B\}$. A^n is defined in the following way:

$$A^0 = \{\lambda\} \quad \text{and} \quad A^{n+1} = AA^n.$$

The *Kleene closure* (or just closure) of A, denoted A^*, is the set

$$A^* = \bigcup_{i=0}^{\infty} A^i.$$

If $a \in \Sigma$ is a single letter, then $\{a\}^*$ will be denoted by a^*. We use the term "class" to denote a set whose elements are languages. Given a class $\mathcal{C}$ of languages over Σ, we define its *class of complements*, denoted co-$\mathcal{C}$, by co-$\mathcal{C} = \{A^c \mid A \in \mathcal{C}\}$.

A *tally* language is a language over a one-letter alphabet. A *bounded* language is a language of the form $A \subseteq a_1^* a_2^* \ldots a_k^*$, where each a_i is a distinct symbol.

2.1. Turing Machines

We shall consider the Turing machine model with a two-way, read-only input tape and a separate two-way, read-write work tape. This model was introduced in (Stearns et. al 1965) to study computations requiring less than linear space. First, we shall describe a nondeterministic Turing machine which consists of:

- A finite *set of states* also called the *finite control* of the machine;
- one semi-infinite *work tape* divided into *cells* (with a leftmost cell and infinite to the right) equipped with a *work head* which can move right or left, and can read or write; and

- an extra tape, called the *input tape*, with *input head* that can move right or left and read only.

A nondeterministic Turing machine is a device whose next move can be chosen from several possibilities. At each moment, the machine is in one of the states. Then it can read the contents of the scanned cells of both the input and work tape, change the contents of the scanned cell of the work tape by writing a new symbol on it, move each head right or left, and change its state. All these operations form a *step*, and are chosen from the possibilities defined by a *transition function*, as a function of the current state and symbols read from the tapes. Observe that the Turing machine is not allowed to change the contents of the input tape.

A machine M starts operating on an *input* $w \in \Sigma^*$ with the input tape containing ¢w\$, where ¢ and \$ $\notin \Sigma$ are special symbols called *left* and *right* *endmarkers*. The work tape is empty, i.e. every cell of the work tape contains a fixed symbol called *blank*. The finite control is in an *initial*, state q_0, the input head scans the first letter of w and the work head scans the first cell of the work tape. Then M proceeds by choosing the next move according to the transition function as long as possible. During the computation the input head cannot move beyond the endmarkers. If there is a sequence of choices that lead to an *accepting state*, then M *accepts* w. Let us now present a more formal definition.

Definition 2.1.1. *A nondeterministic Turing machine is a six-tuple*

$$M = \langle Q, \Sigma, \Gamma, \delta, q_0, F \rangle,$$

where

1. *Q is a finite set of states;*
2. *Σ is an input tape alphabet;*
3. *Γ is a work tape alphabet which contains the blank symbol $B \in \Gamma$;*
4. *$q_0 \in Q$ is an initial state;*
5. *F is a set of accepting states $F \subseteq Q$, and*
6. *$\delta : Q \times (\Sigma \cup \{\text{¢}, \$\}) \times \Gamma \to \mathcal{P}(\{\Gamma - B\} \times Q \times \{R, N, L\}^2)$ is a function called the transition function ($\mathcal{P}(A)$ denotes the set of all subsets of A).*

If the machine is in a state q scanning σ on the input tape and γ on the work tape, and if

$$(\gamma', q', m_1, m_2) \in \delta(q, \sigma, \gamma),$$

then M can in one step change its state to q', print γ' on the work tape, move the input head in the direction m_1, and work head in the direction m_2; $m_i = R$ means "move one cell to the right", $m_i = N$ means "do not move", $m_i = L$ means "move one cell to the left", if there are cells to the left, otherwise "do not move". If $\sigma = \text{¢}$, then $m_1 \neq L$, and if $\sigma = \$$, then $m_1 \neq R$, which means that the input head cannot go beyond endmarkers. Note that the work head cannot write the blank symbol, so cells of the work tape once visited by the head remains non-blank for ever.

A nondeterministic Turing machine is a *deterministic Turing machine* if the range of its transition function δ contains only singleton sets or the empty set. In other words, for every triple (q, σ, γ), there is at most one (γ', q', m_1, m_2) in $\delta(q, \sigma, \gamma)$.

In this paper we consider Turing machines with a single work tape. This is not a serious constraint because any Turing machine with k work tapes can be modified to operate with one work tape and still accept the same language using the same amount of cells on work tape (Hopcroft and Ullman 1979; Balcázar et. al. 1988).

We shall also consider *one-way* Turing machines whose input head cannot move left. In this case an input is read once from left to right. However, it can happen that the input head stays in one cell while the Turing machine changes its state or the contents of the work tape.

2.2. Configuration and Computation

Given a Turing machine M and an input w, a configuration of M on w is a description of the whole status of the computation: it includes the state of the finite control, the position of the input head, the contents of the work tape and position of the work head. More formally.

Definition 2.2.1. *A configuration of a Turing machine M on an input w is a tuple*

$$(q, i, x, j),$$

where

- *q is the current state of M,*
- *i is the position of the input head, $0 \leq i \leq |w| + 1$; if $i = 0$ or $i = |w| + 1$, then the input head scans an endmarker.*
- *$x \in (\Gamma - \{B\})^*$ is the content of the non-blank sector of the work tape, and*
- *j is the position of the work head, $1 \leq j \leq |x| + 1$ (Recall that the work head cannot write the blank symbol).*

Note that the contents of the input tape cannot change during the computation of a machine on a given input. Therefore, there is no need to include the contents of the input tape in the configuration: just the position of the input head is sufficient.

At the beginning of a computation, the machine is in an initial configuration. The *initial configuration* of a machine M on an input w is $(q_0, 1, \lambda, 1)$. A configuration $\beta' = (q', i', x', j')$ can be reached in one step from a configuration $\beta = (q, i, x, j)$ (we also say that β' is an *immediate successor* of β) if and only if there exist $\gamma \in \Gamma$ and $m_1, m_2 \in \{R, N, L\}$ such that:

$$(\gamma, q', m_1, m_2) \in \delta(q, w(i), x(j)),$$

$$x'(k) = \begin{cases} x(k) & \text{for } k \neq j, \\ \gamma & \text{for } k = j, \end{cases}$$

$$i' = i + ||m_1||, \qquad j' = j + ||m_2||, \quad \text{and} \quad j' \geq 1,$$

where $||R|| = 1$, $||N|| = 0$, $||L|| = -1$. Recall that $w(i)$ denotes the i-th letter of w, and by convention, $w(0) = \mathbf{\phi}$, $w(|w| + 1) = \$$, and $x(|x| + 1) = B$.

Note that the transition function does not allow $i' < 0$ or $i' > |w| + 1$, so the input head cannot go beyond the endmarkers. But it may happen that $j = 1$ and $m_2 = L$ (the work head standing in the first cell tries to go left). In that case, the condition $j' \geq 1$ does not allow the work head to go left.

A *final configuration* is a configuration which does not have any immediate successors. An *accepting configuration* is a final configuration with an accepting state.

A *computation* of a machine M on an input w is a (finite or infinite) sequence of configurations of M

$$\beta_0, \beta_1, \beta_2, \ldots$$

such that β_0 is the initial configuration and each β_i is an immediate successor of β_{i-1}. An *accepting computation* is a finite computation ending with an accepting configuration. The input w is *accepted* by M if and only if there exists an accepting computation of M on w. The *language accepted* by the machine M, denoted $L(M)$, is the set of all words accepted by M.

An *accessible configuration* is a configuration that can be reached from the initial configuration by a finite computation. The set of all accessible configurations of a nondeterministic Turing machine M on an input w form a *tree of accessible configurations* in the following manner: the nodes are labeled by configurations of M on w; the root is (labeled by) the initial configuration and for any node labeled by a configuration β, its sons are (labeled by) the immediate successors of β. Leaves of the tree are labeled by final configurations, some of which may be accepting. Note that two nodes of the tree can be labeled by the same configuration and that the tree can be infinite. The input is accepted if and only if there is at least one accepting leaf (labeled by an accepting configuration) in the tree. In that case there is a finite path leading from the root to the accepting leaf and the labels on the path form an accepting computation.

Note that for a deterministic Turing machine, every move is completely determined by the current configuration. The state of the machine and symbols scanned by the tape heads completely determine the next state and the moves of the heads. Hence, there is exactly one computation of the deterministic Turing machine on every input w, and the tree of accessible configurations is in this case reduced to a single path.

We shall also consider Turing machine which recognize languages. An *recognizing* deterministic or nondeterministic Turing machine is like normal Turing machine but it is equipped with two special states (accepting) s_a and (rejecting) s_r. And for any input w exactly one of the following can happen: either the machine reaches the state s_a (and then the input is accepted) or it reaches the state s_r. Note that a recognizing deterministic Turing machine stops on every input; in s_a or s_r. On the other hand, there can be many computations (some of them even infinite) of recognizing nondeterministic Turing machine, on both accepted and not accepted inputs. We only require that for every accepted input there is at least one finite accepting computation (ending with s_a) and for every not accepted input there is at least one finite rejecting (ending with s_r) computation; and it can not happen that on one input both s_a and s_r can be reached.

Roughly speaking, when a recognizing Turing machine reaches the state s_a, then it knows that the input is accepted; and when it reaches the state s_r, it knows that the input is not accepted.

The following lemma gives the upper bound for the number of all configurations.

Lemma 2.2.2. Suppose that a Turing machine M has s states, a work tape alphabet of t symbols, and which on an input of length n uses no more than k cells on the work tape. Then M cannot enter more than $s(n+2)kt^k$ different configurations.

The factor s represents the number of possible states of M, $(n+2)$ the possible input head positions on the input tape (including endmarkers), k the possible head positions on the work tape, and t^k the possible patterns which can be written on the work tape.

2.3. Internal Configuration

An *internal configuration* of a Turing machine M is a description of the "internal status" of the computation.

Definition 2.3.1. *An internal configuration of M is a tuple*

$$(q, x, j),$$

where

- *q is the current state of M,*
- *$x \in (\Gamma - \{B\})^*$ is the contents of the non-blank sector of the work tape, and*
- *j is the position of the work head $1 \leq j \leq |x| + 1$.*

The following lemma gives the upper bound for the number of all internal configurations.

Lemma 2.3.2. Suppose that a Turing machine M has s states, a work tape alphabet of t symbols, and uses no more than k work tape cells on an input. Then M cannot enter more than skt^k different internal configurations. And there is a constant d, depending on s and t only, such that $d^k \geq skt^k$.

The factor s represents the number of possible states of M, k the possible head positions on the work tape, and t^k the possible patterns which can be written on the work tape.

2.4. Alternating Turing Machines

The concept of the *alternating Turing machine* was introduced in (Chandra et. al. 1981) as a generalization of the nondeterministic Turing machine. An alternating Turing machines works exactly like the nondeterministic Turing machine, only the definition of acceptance is different. In the case of nondeterministic Turing machines, the transition function allows a single configuration α to reach several configurations $\beta_1, \ldots, \beta_k$ in one step; by definition, the configuration α leads to acceptance if and only if there exists a successor β_i which leads to acceptance. In addition to these *existential branches*, an alternating Turing machine can also make *universal branches*. In a universal branch, a configuration α can again reach several configurations $\beta_1, \ldots, \beta_k$ in one step, but now α leads to acceptance if and only if all successors $\beta_1, \ldots, \beta_k$ lead to acceptance.

More precisely, all states of the alternating Turing machine are partitioned into *universal* and *existential* states. A configuration is called universal or existential if the state associated with the configuration is universal or existential, respectively. A *computation tree* of an alternating Turing machine M on an input w is a tree whose nodes are labeled by configurations of M on w. The root of the tree is labeled by the initial configuration; the children of any node labeled by a universal configuration are all immediate successors of that configuration; and any node labeled by an existential configuration has one child, which is labeled by one of the immediate successors of that configuration (provided there are any). An *accepting tree* is a finite computation tree such that all leaves of the tree are accepting configurations. M accepts w if there is an accepting computation tree of M on w.

Note that a nondeterministic Turing machine is an alternating Turing machine with existential states only.

Now we define an alternating Turing machines with a bounded number of alternations between universal and existential states on every computation.

Definition 2.4.1. *Let M be an alternating Turing machine. We say that M is k alternation-bounded if whenever there is a sequence of configurations on any input w*

$$\beta_0, \beta_1, \beta_2, \ldots, \beta_m$$

such that for every i, $0 \le i \le m - 1$, β_{i+1} is reachable from β_i and β_i is a universal configuration if and only if β_{i+1} is existential, then $m < k$.

In other words, M is k alternation-bounded if any computation alternates universal and existential configurations at most $k - 1$ times.

Definition 2.4.2. *For every $k \geq 1$, a Π_k-alternating (or Σ_k-alternating) Turing machine is a k alternation-bounded Turing machine such that the initial state is universal (or existential, respectively).*

Note that a Σ_1-alternating Turing machine is a nondeterministic Turing machine. By convention, a Σ_0- or Π_0-alternating Turing machine is a deterministic Turing machine.

2.5. Space Complexity

The *space* used by the configuration (q, i, x, j) is $|x|$ –the number of non-blank cells on the work tape. The space used by a computation (or a computation tree for alternating Turing machine) is the maximal space used by configurations of the computation (or the tree). Let us recall that every cell that has ever been visited by the work head is non-blank. This assumption has been made to simplify the definition of space complexity. In another definition which has been used in literature, Turing machines are allowed to write the blank symbol on the work tape and then space used by a computation is defined as the number of cells that has ever been visited by the work head. These definitions are equivalent because every Turing machine that can write blanks can be simulated step by step by a Turing machine that cannot write blanks, but uses an additional symbol B' that is written instead of the blank and is not distinguished from the blank when reading.

Now we define space-bounded Turing machines. Let $L(n)$ be a function on natural numbers.

Definition 2.5.1. *A Turing machine is said to be strongly $L(n)$ space-bounded if no accessible configuration on any input of length n, uses more than $L(n)$ space.*

Definition 2.5.2. *A Turing machine is said to be weakly $L(n)$ space-bounded if for every accepted input w, at least one accepting computation tree (or computation for nondeterministic Turing machine) uses at most $L(|w|)$ space.*

We also define the middle mode of space complexity.

Definition 2.5.3. *A Turing machine is said to be middle $L(n)$ space-bounded if no accessible configuration on any accepted input of length n, uses more than $L(n)$ space.*

Note that the definition of middle mode of space complexity only requires that accepted inputs do not lead to overuse of space. It should be obvious that every Turing machine uses at least one cell on all inputs, so if $L(n)$ bounds space, then $L(n) \geq 1$ for all n. We shall use the convention that "$L(n)$ space-bounded" really means "$\max(1, L(n))$ space-bounded".

$DSPACE[L(n)]$, $NSPACE[L(n)]$, or $ASPACE[L(n)]$ denotes the class of languages accepted by deterministic, nondeterministic, or alternating $L(n)$ space-bounded Turing machines, respectively. These classes are called *space complexity classes*, deterministic, nondeterministic, or alternating, respectively. Π_k-$SPACE[L(n)]$ and Σ_k-$SPACE[L(n)]$ denotes the class of languages accepted by $L(n)$ space-bounded Π_k- or Σ_k-alternating Turing machines, respectively. We add the prefix *strong*, *weak*, or *middle* if we consider classes accepted by strongly, weakly, or middle space-bounded Turing machines, respectively. We shall omit the prefix if it does not matter which mode is used. We add the prefix *one-way* if we consider classes accepted by one-way Turing machines.

Our Turing machines are allowed an arbitrary finite work tape alphabet. Thus, any Turing machine that operates in space $L(n)$ can be modified to operate in space $cL(n)$ for any constant $c > 0$, and still accept the same language (Hopcroft and Ullman 1979; Balcázar et. al. 1988). Hence, it suffices to consider only growth rates of functions which bound space. Let $f(n)$ and $g(n)$ be two functions. We shall use the following notations:

- $f(n) = O(g(n))$ if there exists a constant $c > 0$ such that $f(n) \leq cg(n)$,
- $f(n) = o(g(n))$ if $\lim_{n \to \infty} \frac{f(n)}{g(n)} = 0$,
- $f(n) \ll g(n)$ if $\liminf_{n \to \infty} \frac{f(n)}{g(n)} = 0$.

Note that if $f(n) = o(g(n))$, then $f(n) \ll g(n)$ but not necessarily otherwise. For example, if $g(n) = \log n$ and

$$L(n) = \begin{cases} 1, & \text{for } n \text{ even,} \\ \log n, & \text{otherwise.} \end{cases}$$

then $f(n) \ll g(n)$ but not $f(n) = o(g(n))$. Note also that if $f(n) \ll g(n)$ or $f(n) = o(g(n))$, then there is no constant $c > 0$ such that $f(n) \geq cg(n)$.

The logarithm function $\log n$ is in base 2. The functions $\log^{(k)} n$ and $\log^* n$ are defined in the following way:

$$\log^{(0)} n = n$$
$$\log^{(k)} n = \log(\log^{(k-1)} n) \qquad \text{for } k \geq 1$$
$$\log^* n = \min\{i \mid \log^{(i)} n \leq 1\}.$$

3. Languages Acceptable with Logarithmic Space

Before we start to discuss the property of languages acceptable with sublogarithmic space we first present, in this chapter, some examples of languages acceptable with logarithmic space and we show that the Turing machines with logarithmic or greater space can store on the work tape numbers up to the size of the input, and they can remember any position on the input tape. As we shall see in the sequel, the one of the greatest disadvantages of Turing machines with sublogarithmic space is that they cannot store such great numbers and remember an arbitrary position of the input tape. In Section 3.2 we show that Turing machines with logarithmic space are equivalent to pebble automata. And in Section 3.3 we present the language GAP (graph accessibility problem) acceptable by a $\log n$ space-bounded nondeterministic Turing machine and we prove that GAP belongs to $DSPACE(\log n)$ if and only if $DSPACE(\log n) = NSPACE(\log n)$. In other words we prove that GAP is complete for the class $NSPACE(\log n)$.

3.1. Two Examples of Languages Acceptable with Logarithmic Space

The first example of a language acceptable in logarithmic space is

$$\{a^n \mid n \text{ is prime}\}.$$

A strongly $\log n$ space-bounded deterministic Turing machine M which accepts the language first counts the letters of an input. To do this, M goes through the input and counts its steps. The successive numbers are stored in binary on the work tape. Next, M checks one by one for each $k < n$, if k divides n; and accepts if n is not divisible by any $k < n$.

The second example of a language acceptable in logarithmic space is

$$\{ww^R \mid w \in \{0, 1\}^*\},$$

where w^R is the reflection of w (w written from right to left). A strongly $\log n$ space-bounded deterministic Turing machine M accepting the language first compares the first letter of the input with the last one, then the second with the last by one, and so on. The machine M keeps in binary on the work tape the position of the letter just compared.

We can see from these examples that the Turing machines with logarithmic or greater space can store on the work tape numbers up to the size of the input, and they can remember any position on the input tape.

3.2. Pebble Automata

In this section we show that Turing machines with logarithmic space are equivalent to the finite automata with pebbles.

Definition 3.2.1. *A k-pebble finite automaton is a machine with a two-way read-only input tape, and k pebbles which can be placed on and removed from the input tape. The action of the k-pebble automaton depends on the current state, the currently scanned input symbol, and the set of the pebbles on the currently scanned cell. The action consists of moving the input head, changing the state of the finite control, and picking up or placing (or none) some pebbles on the currently scanned cell on the input tape.*

One pebble does not increase the power of the finite automaton, for 1-pebble automata accept only regular languages, see Corollary 15.3.6 below. But the nonregular language $\{ww^R \mid w \in \{0,1\}^*\}$ can be accepted by a 2-pebble automaton. The automaton places one pebble at the beginning of the input, the other at the end, compares the letters under pebbles, moves both pebbles one cell to the center and repeats the process until both pebbles meet in the center.

It is easy to see that any k-pebble automaton can be simulated by a $\log n$ space-bounded Turing machine. The Turing machine remembers (in binary) the positions of the input head and the pebbles of the automaton, and simulates step by step its computation. Hence we have proved the following theorem.

Theorem 3.2.2. For every deterministic (or nondeterministic) k-pebble automaton there is a deterministic (or nondeterministic) Turing machine with logarithmic space that simulates the automaton.

But also opposite holds.

Theorem 3.2.3. For every deterministic (or nondeterministic) Turing machine with logarithmic space there exists a deterministic (or nondeterministic, respectively) k-pebble automaton that simulates the Turing machine, where k depends on the number of work tape symbols of the Turing machine.

Proof. We shall prove the theorem in four stages.

In the first stage the work tape of the Turing machine is replaced by two push-down tapes.

A push-down tape is like a work tape of Turing machine, but is used in specific way: Initially the push-down tape contains only a distinguished symbol s_0 of the bottom of the push-down tape in the first cell and the blank symbol in all others. The symbol s_o cannot be changed or used in any other cell. Further, the pushdown head can never shift left unless it first erases (i.e. overwrites the blank symbol) the symbol previously scanned, and it can never shift right from a cell leaving blank symbol. Thus, throughout the computation all cells on the pushdown tape are blank except exactly these between s_o and the pushdown head. There are three operations possible on the push-down tape: add new symbol on the top (the right end) of the push-down tape, check if the push-down tape is empty (i.e., contains only s_o), and take off the symbol from the top (provided the push-down tape is not empty). For more details see (Hopcroft and Ullman 1979).

While simulating the work tape of the Turing machine one push-down tape contains the contents of the work tape to the left from the work head and the other to the right. Symbols which are further from the work head of the Turing machine stay nearer to the bottom of the push-down tapes. Each push-down tape contains no more unblank cells than the simulated work tape and both the push-down tapes are $\log n$ space-bounded.

In the second stage the two $\log n$ space-bounded push-down tapes are simulated by three n^c space-bounded counters, where c is a constant. A counter is a push-down tape with only one symbol (plus s_o). There are only three operations possible on the counter: increase by one, check for zero, decrease by one. Contents of the push-down tapes are treated as numbers in $(r + 1)$-ary, where r is the number of symbols of the pushdown tapes (zero is not used to avoid ambiguity). These two numbers representing contents of the push-down tapes are kept in unary in the first and second counter. The third counter is used to manipulate with these numbers. Putting a new symbol on the top of a pushdown tape is simulated by multiplying the counter by $(r + 1)$ and adding the appropriate number less than $r + 1$. Taking off a symbol is simulated by dividing the counter by $r + 1$ and decoding the remainder. Maximal contents of each counter is $(r+1)^{\log n} \leq n^c$, for some constant c. Multiplication of the content of the first (or second) counter by $(r + 1)$ is done recursively by decreasing the counter by one and increasing the third counter by $(r+1)$ until the first (or second) counter is empty. Division is made in a similar way.

In the third stage the maximal capacity of the counters is reduced to n by increasing the number of counters. We show how a counter with capacity of n^2 can be simulated by two counters with capacity of n. Further reduction, if needed, can be obtained by repetition of this process. Let x and y be the contents of two new counters. They represent the number

$$f(x,y) = y + \sum_{i=0}^{x+y} i$$

on the old counter. The function $f(x, y)$ is the well known pair function: $f(0;0) = 0$, $f(1,0) = 1$, $f(0,1) = 2$, $f(2,0) = 3$, $f(1,1) = 4$, and so on. Simulation of increasing by one the old counter is done by:

$$\text{if} \quad x > 0 \quad \text{then} \quad y \leftarrow y+1; \quad x \leftarrow x-1$$
$$\text{else} \quad x \leftarrow y+1; \quad y \leftarrow 0$$

Decreasing by one is simulated by:

$$\text{if} \quad y > 0 \quad \text{then} \quad y \leftarrow y - 1; \ x \leftarrow x + 1$$
$$\text{else} \quad x \leftarrow 0; \ y \leftarrow x - 1$$

It is easy to see that in that way numbers up to $(n+1)(n+2)/2$ can be kept by two counters with capacity of n. It is also obvious how the counter with the capacity of $c \cdot n$ can be simulated by a counter with the capacity of n, for any positive constant c.

In the fourth stage each counter of capacity n is simulated by the position of one pebble on the input tape.

Hence we have shown that every $\log n$ space-bounded Turing machine can be simulated by a k-pebble automaton, where k depends on the number of work tape symbols of the Turing machine. □

3.3. NSPACE(log n) Complete Language

An important example of languages acceptable by nondeterministic Turing machines with logarithmic space is the language GAP which consists of encoded directed graphs which have a path from the first to last vertex.

A directed graph $G = (V, E)$ consists of a set of vertices $V = \{a_1, a_2, \ldots, a_k\}$, which is a subset of natural numbers, and a set of edges $E \subseteq V \times V$. The graph is encoded as

$$* * a_1 * a_{1,1} * a_{1,2} \ldots a_{1,i_1} * * a_2 * a_{2,1} * a_{2,2} \ldots a_{2,i_2} * * \ldots$$

where $a_{1,1} * a_{1,2} \ldots a_{1,i_1}$ is the list of vertices reachable by an edge from the vertex a_1, and $a_{2,1} * a_{2,2} \ldots a_{2,i_2} * * \ldots$ are vertices reachable from the vertex a_2, and so on. First we show that

Lemma 3.3.1.

$$GAP \in NSPACE(\log n).$$

Proof. A nondeterministic Turing machine M guesses the path vertex by vertex. It starts from the vertex a_1 written as $* * a_1 *$ on the input tape, chooses one vertex from the list $a_{1,1} * a_{1,2} \ldots a_{1,i_1}$, say $a_{1,1}$, goes to the place where $* * a_{1,1} *$ is written and repeats the process until it reaches the vertex a_k. The machine M does not write $a_{1,1}$ on the work tape, for it can be too large. Instead, it remembers the position of $* a_{1,1} *$ on the input tape and looks for $* * a_{1,1} *$ through all places of the form $* * x *$. To check whether $x = a_{1,1}$, M compares x and $a_{1,1}$ digit by digit. □

It is not known whether GAP belongs to the class $DSPACE(\log n)$. But it is known that if GAP belongs to the class $DSPACE(\log n)$ then $NSPACE(\log n)$ = $DSPACE(\log n)$. In order to prove this fact, we shall need few definitions and lemmas.

Definition 3.3.2. *A log-space transducer is a deterministic Turing machine which have a read-only input tape, write-read work tape, and the write-only output tape on which the head never moves left.*

Definition 3.3.3. *We say that a language A_1 is log-space reducible to a language A_2 if there is a log-space transducer that for every input x produces an output y that is in A_2 if and only if x is in A_1, and uses at most $\log|x|$ space on the work tape.*

Definition 3.3.4. *A language $A \in NSPACE(\log n)$ is called $NSPACE(\log n)$ complete if every language B from $NSPACE(\log n)$ is log-space reducible to A.*

Lemma 3.3.5. The language GAP is $NSPACE(\log n)$ complete.

Proof. By Lemma 3.3.1, $GAP \in NSPACE(\log n)$, so it suffice to prove that any language $B \in NSPACE(\log n)$ is log-space reducible to GAP. Let M be a log space-bounded nondeterministic Turing machine accepting B. We construct a log-space transducer M_1 which reduces B to GAP as follows. For any input x, M_1 produces the graph G_x whose vertices are configurations of M on x and edges join every configuration with its immediate successors; the first vertex is the initial configuration, and the last the accepting configuration (we can assume that there is only one accepting configuration). M_1 writes on its output tape one by one the configurations of M on x along with the lists of immediate successors of any configuration. It is easy to see that the computation can be done within $\log|x|$ space and that $x \in L$ if and only if $G_x \in GAP$.   □

Lemma 3.3.6. If A_1 is log-space reducible to A_2 and $A_2 \in DSPACE(\log n)$ then $A_1 \in DSPACE(\log n)$.

Proof. Let M_1 be a log-space transducer which reduces A_1 to A_2 and let M_2 be a log space-bounded deterministic Turing machine accepting A_2. On an input x of length n, M_1 produces an output y of length bounded by n^c for some constant c. Note that y cannot be written in $\log n$ space.

We construct a $\log n$ space-bounded deterministic Turing machine M_3 accepting A_1 as follows. M_3 simulates M_2 working on y. M_3 remembers only the position of the input head of M_2 in base 2^c. Since the input position is less than n^c, this number can be stored in $\log n$ space. Each time the i-th symbol of y is needed it is computed by simulating M_1 on x.   □

Using Lemma 3.3.5 and Lemma 3.3.6 one can easily prove the following theorem.

Theorem 3.3.7.

$$GAP \in DSPACE(\log n)$$

if and only if

$$NSPACE(\log n) = DSPACE(\log n).$$

4. Examples of Languages Acceptable with Sublogarithmic Space

In this chapter we present some examples of languages that are accepted with sublogarithmic space. It is a well known fact that every regular language can be accepted by a Turing machine with zero space and every Turing machine with space bounded by a constant accepts only a regular language (Hopcroft and Ullman 1979). It is also easy to show that there exist non-regular languages accepted in space $L(n) \ll \log n$ (i.e. $\liminf \frac{L(n)}{\log n} = 0$). For example, the language

$$\{a^n \mid n \text{ is prime}\}$$

can be accepted in the space

$$L(n) = \begin{cases} 0, & \text{for } n \text{ even}, \\ \log n, & \text{otherwise}. \end{cases}$$

A strongly $L(n)$ space-bounded deterministic Turing machine M which accepts the language first checks, without using any space, whether n is even or not. If n is even, then M accepts only if $n = 2$. If n is odd, then M counts the letters of the input and checks one by one for each $k < n$, if k divides n; M accepts the input if n is not divisible by any $k < n$.

4.1. Languages Acceptable with o(log n) Space

The first example of a non-regular language accepted in $o(\log n)$ space was presented by Stearns et. al. (1965). Their language consists of all words over the alphabet $\{0, 1, \#\}$ of the form

$$w_k = b_0 \# b_1 \# \ldots \# b_k,$$

where b_i is the binary description of the number i. A strongly $\log \log n$ space-bounded Turing machine M that accepts the language checks one by one if every number b_i is by one greater than its predecessor b_{i-1}. In order to do this, M while going from b_i to b_{i-1} compares one by one the digits of both numbers. It has to remember only the digit which is compared and its position in the binary description of the number. Thus, M uses exactly $\lfloor \log \lfloor \log k \rfloor \rfloor + 1$ space if the input is of the form $w_k x$ with $w_k = b_0 \# b_1 \# \ldots \# b_k$, and $x \in \{0, 1, \#\}^*$ that does not begin with $\# b_{k+1} \#$ (M does not use any space if the input does not begin with $0\#$). Hence, M is strongly $\log \log n$ space-bounded.

To present further examples, we shall need the following two lemmas.

Lemma 4.1.1. (Pumping lemma for regular languages) Let R be a regular language. Then there is a constant r such that any word $z \in R$ longer than r may be decomposed into $z = uvw$ in such a way that $|uv| \leq r$, $|v| \geq 1$, and $uv^i w \in R$ for all $i \geq 0$.
Furthermore, r can be taken to be no more than the number of states of any finite automaton accepting R.

Lemma 4.1.2 Consider the following three functions:

$G(k) = lcm\{j \mid j \leq k\}$,
$F(k) = \min\{i \mid i \text{ does not divide } k\}$, and
$H(k) = \max\{F(i) \mid i \leq k\}$,

where $lcm\ A$ denotes the least common multiple of all numbers from A. Then:

(a)
$$G(k) = \prod_{all\ primes\ p} p^{\lfloor \log_p k \rfloor},$$

Observe that $p^{\lfloor \log_p k \rfloor} = 1$ for all $p > k$.
(b) $c_1^k \leq G(k) \leq c_2^k$ for every $k \geq 2$ and some constants $c_1, c_2 > 1$,
(c) if m is an arbitrary common multiple of all $j \leq k$ and $k \geq 2$, then $m \geq c_1^k$.
(d) $F(k) \leq c \cdot \log k$ for for every $k \geq 2$ and some constant c.
(e) there exists a constant c such that for every pair of natural numbers k and m, $k \neq m$, there exists a number $j < c \cdot \log |k + m|$ such that $k \neq m \pmod{j}$.
(f) $F(G(k)) > k$,
(g) $d \log k \leq H(k) \leq c \log k$ for every $k \geq 2$ and some constants $c, d > 0$,

Proof. (a) Follows immediately from (Narkiewicz 1983, Thm 1.8(ii)).
(b) Let $k \geq 2$. It is obvious that for all $p \leq k$,

$$p \leq p^{\lfloor \log_p k \rfloor} \leq k < p^{\lfloor \log_p k \rfloor + 1}.$$

Note that also
$$\sqrt{k} < p^{\lfloor \log_p k \rfloor},$$

since otherwise $\sqrt{k} \geq p^{\lfloor \log_p k \rfloor} \geq p$ hence, $k \geq p^{\lfloor \log_p k \rfloor + 1}$, a contradiction.
So, we have

$$k^{\pi(k)/2} = \prod_{all\ primes\ p} \sqrt{k} \leq G(k) \leq \prod_{all\ primes\ p} k = k^{\pi(k)},$$

where $\pi(k)$ denotes the number of all primes not greater than k. By the Čebyšev's (Tchebychev's) theorem (Narkiewicz 1983, Thm 3.1), there exist two positive constants d_1, d_2, such that

$$d_1 \frac{k}{\log k} \leq \pi(k) \leq d_2 \frac{k}{\log k}.$$

Hence,

$$2^{d_1 k/2} = k^{d_1 \frac{k}{2\log k}} \leq lcm\{j \mid j \leq k\} \leq k^{d_2 \frac{k}{\log k}} = 2^{d_2 k}.$$

Thus, if we put $c_1 = 2^{d_1/2}$ and $c_2 = 2^{d_2}$, we obtain (b).

(c) Follows immediately from (b).

(d) Observe that k is a common multiple of all $i \leq F(k) - 1$, and by (c), $k \geq c_1^{F(k)-1}$. Taking logarithm we obtain $F(k) \leq c \log k$ for some constant $c > 0$.

(e) Let

$$j = F(|k - m|).$$

Then $j = 2$ if $|k - m| = 1$, and by (d) $j < c \cdot \log |k - m|$ if $|k - m| > 1$. Hence, there exists a constant $c > 0$ such that $j < c \cdot \log |k + m|$.

Furthermore, if $k \neq m$, then j does not divide $|k - m|$, so $k \neq m \,(\mathrm{mod}\, j)$.

(f) follows directly from the definitions.

(g) By (d), $F(s) \leq c \log k$, for every $s \leq k$, so $H(k) \leq c \log k$.

Let $m = \max\{j \mid G(j) \leq k\}$. Then $G(m) \leq k < G(m + 1)$. Hence, by (b), $k \leq c_2^{m+1}$ and $m \geq d \log k$, for some constant $d > 0$. Thus, $H(k) \geq F(G(m)) > m \geq d \log k$.   □

Now we can present further examples of languages acceptable with sublogarithmic space (Freivalds 1975).

Lemma 4.1.3.   Let

$$A = \{a^k b^m \mid k \neq m\}.$$

Then

- A is non-regular,
- $A \in weak\text{-}DSPACE[\log \log n]$,
- $A \in weak\text{-}one\text{-}way\text{-}NSPACE[\log \log n]$.

Proof.   First we show that A can be accepted by a weakly $\log \log n$ space-bounded one-way nondeterministic Turing machine M. On any input of the form $a^k b^m$, M guesses a number j such that $k \neq m(\mathrm{mod}\, j)$. The work tape of M is divided into three tracks; j is written in binary on the first track. The second and third tracks are used as counters modulo j to count a's and b's. The content of both counters is written in binary on the tracks. M starts at the beginning of the input with empty counters. Then it counts a's modulo j by going right and increasing the first counter after each step. When M reaches the first b, the second track contains $k(\mathrm{mod}\, j)$ and then M counts b's modulo j on the third track. At the end, M compares the contents of the second and third tracks. M is weakly $\log \log n$ space-bounded because if an input of the form $w = a^k b^m$ is in A, then $k \neq m$ and by Lemma 4.1.2(e), there is j such that $k \neq m(\mathrm{mod}\, j)$ and $j < c \log |k + m|$. Hence, there is an accepting computation of M using $d \log \log |w|$ space for some constant $d > 0$.

The language A can also be accepted by a weakly $\log\log n$ space-bounded two-way deterministic Turing machine which instead of guessing j tries one by one all successive natural numbers, starting from 1, up to the moment it finds a suitable one.

To see that A is non-regular, consider the language

$$B = \{a^k b^k \mid k \geq 0\} = A^c \cap a^* b^*,$$

where A^c is the complement of A. In the next lemma we shall show that B is non-regular. Hence, A is also non-regular.   □

Lemma 4.1.4.   Let
$$B = \{a^k b^k \mid k \geq 0\}.$$

Then B is non-regular and, moreover, B does not contain any infinite regular subset.

Proof.   Suppose, for a contradiction, that B contains some infinite regular subset R that is accepted by a finite automaton with s states. Since R is infinite, there is k such that $k > s$ and $a^k b^k \in R$. It is easy to see that $a^k b^k$ cannot be decomposed into $a^k b^k = uvw$ in such a way that $|uv| < s$, $|v| \geq 1$, and $uv^i w \in R \subseteq B$ for every $i \geq 0$. Thus, by Lemma 4.1.1, R cannot be regular.   □

Note that the Turing machines described in the proof of Lemma 4.1.3 accept the language A in the weak mode but not in the strong mode, since they possibly use too much space if an input is not in A. In Chapter 8 we shall show that A cannot be accepted by any strongly $o(\log n)$ space-bounded deterministic, nondeterministic, or alternating Turing machine.

4.2. Tally Languages Acceptable with Sublogarithmic Space

In this section we give an example of a tally language (i.e. a language over a one-letter alphabet) that can be accepted by strongly $o(\log n)$ space-bounded deterministic Turing machines. In the following lemma we consider a language that was presented in (Alt and Melhorn 1975).

Lemma 4.2.1.   Let

$$C = \{a^n \mid F(n) \text{ is a power of } 2\},$$

where
$$F(n) = \min\{i \mid i \text{ does not divide } n\}.$$

Then

- C is non-regular and

- $C \in strong\text{-}DSPACE[\log\log n]$.

Proof. First we present a strongly $\log\log n$ space-bounded deterministic Turing machine M accepting C. On an input of the form $w = a^n$, M computes $F(n)$ in binary and checks if $F(n)$ is a power of 2. To compute $F(n)$, M writes in binary on its work tape one by one successive natural numbers up to the moment it finds one that does not divide n. In order to check if a number k written in binary on the work tape divides n, M forms a counter modulo k on the work tape and going from one end of the input to the other counts a's. By Lemma 4.1.2(d), $F(n) \le c\log n$, so M needs only $O(\log\log n)$ space to remember it in binary. It is easy to check if the number written in binary is a power of 2. Hence, M is strongly $\log\log n$ space-bounded.

Now we shall show that C is non-regular. Assume that C can be accepted by a finite automaton with s states. Let r be such a number that $2^r > s$ and let

$$n = lcm\{i \mid i \le 2^r - 1\},$$

where $lcm\ A$ denotes the least common multiple of all numbers from A.

By Lemma 4.1.2(f), $F(n) \ge 2^r$ and by Lemma 4.1.2(a), n is divisible by 2^{r-1} and not by 2^r, so $F(n) = 2^r$ and $a^n \in C$. By the pumping lemma 4.1.1, a^n can be decomposed into $a^n = uvw$ with $|uv| \le s$, $|v| \ge 1$, such that $uv^{j+1}w \in C$, where $j = \frac{n}{|v|}$ ($|v|$ divides n, since $|v| \le s < 2^r$). But

$$|uv^{j+1}w| = |uvw| + j|v| = n + n = 2n.$$

Note that
$$2^r < F(2n),$$
since every $i \le 2^r$ divides $2n$. On the other hand,
$$F(2n) < 2^{r+1},$$
since there is a prime number p such that $2^r < p < 2 \cdot 2^r = 2^{r+1}$ and p divides neither n nor 2, so $F(2n) \le p$.

Hence, $F(2n)$ is not a power of 2 and $uv^{j+1}w \notin C$, a contradiction. We conclude that C is non-regular.   □

5. Lower Bounds for Accepting Non-regular Languages

In this chapter we present so-called gap theorems. It is impossible for a Turing machine to lay off small amounts of tape; either it lays off an amount bounded by a constant or the amount laid off must reach $d \log \log n$ infinitely often for some constant d. The first results of this type were presented in (Stearns et. al 1965) for strongly space-bounded deterministic Turing machines which stop on every input. Hopcroft and Ullman (1967) showed that the proofs of (Stearns et. al 1965) can also be used for nondeterministic Turing machines and they removed the halting restriction. Alberts (1985) showed that $\log \log n$ is the lower bound for accepting non-regular languages by weakly space-bounded deterministic and nondeterministic Turing machines. Iwama (1993) proved that $\log \log n$ is the lower bound for weakly space-bounded alternating Turing machines. In all cases, the lower bound $\log \log n$ is tight because, as we showed in Chapter 4, there exists a non-regular language accepted by a strongly $\log \log n$ space-bounded deterministic Turing machine.

We shall also consider lower bounds for accepting non-regular languages by one-way Turing machines, i.e. the Turing machines whose input head can move only from left to right. Stearns et. al. (1965) showed that the lower bound for strongly space-bounded one-way deterministic Turing machines is $\log n$ and their proof can easily be adopted to prove that $\log n$ is the lower bound also for strongly space-bounded nondeterministic and alternating Turing machines. The lower bound for weakly space-bounded one-way Turing machines is equal to $\log n$ for deterministic, and to $\log \log n$ for nondeterministic and alternating Turing machines (Alberts 1985). Both $\log n$ and $\log \log n$ lower bounds are tight because by Lemma 4.1.3, the language $A = \{a^k b^m \mid k \neq m\}$ can be accepted by a weakly $\log \log n$ space-bounded one-way nondeterministic Turing machine and it is easy to see that A can be accepted by a strongly $\log n$ space-bounded one-way deterministic Turing machine.

We shall also consider lower bounds for a middle mode of space complexity, which requires that all computations on every accepted input satisfy the space bounds. For two-way Turing machines, $\log \log n$ is the lower bound for accepting non-regular languages for both the weak and strong mode of space complexity. Hence, $\log \log n$ is also the lower bound for middle space-bounded two-way Turing machines. The situation is more interesting for one-way Turing machines. In (Szepietowski 1988) it was shown that for one-way Turing machines the lower bounds for the middle mode are different from both the lower bounds for the weak and strong mode. The lower bound for the middle space-bounded one-way

Turing machines is equal to $\log \log n$ for alternating, and to $\log n$ for deterministic and nondeterministic Turing machines. These lower bounds are tight.

In Chapter 15 we shall show that $\log \log n$ is the lower bound for accepting non-regular languages by 1-pebble (and 1-inkdot) Turing machines, and that Turing machines with two-dimensional inputs can be space-bounded by functions that grow much slower, e.g. $\log^* n$ or $\log^{(k)} n$ for every k.

5.1. Lower Bounds for Two-way Turing Machines

In this section we show that the lower space bound for accepting non-regular languages is equal to $\log \log n$ or, more precisely, that every (deterministic, non-deterministic, or alternating) $o(\log \log n)$ space-bounded Turing machine accepts a regular language and actually uses a bounded amount of space on every input. We shall prove this for the weak mode of space complexity, but since every strongly space-bounded Turing machine is also weakly space-bounded, the same is true for the strong mode of space complexity. The lower bound $\log \log n$ is tight because, as we showed in Chapter 4, there exists a non-regular language accepted by a strongly $\log \log n$ space-bounded deterministic Turing machine.

First, we show that $\log \log n$ is the lower bound for nondeterministic (or deterministic) Turing machines. Let M be a nondeterministic Turing machine with an input alphabet Σ. Let k be a natural number and w a word over Σ. Let C_M^k denote the set of all k space-bounded internal configurations of M. Consider the following two relations.

- $P_M^k(w) \subseteq C_M^k \times C_M^k$
  $(\beta_1, \beta_2) \in P_M^k(w)$ if and only if there is an k space-bounded computation of M that starting in β_1 at the first letter of w reaches β_2 just after it leaves w to the left.
- $Q_M^k(w) \subseteq C_M^k$
  $\beta \in Q_M^k(w)$ if and only if there is an k space-bounded accepting computation of M that starting in β at the leftmost letter of w accepts without leaving w.

Now we shall define an equivalence relation.

Definition 5.1.1. *Given $u, v \in \Sigma^*$, u and v are called k-equivalent, denoted $u \equiv_k v$, if and only if*

$$P_M^k(u) = P_M^k(v) \quad \text{and} \quad Q_M^k(u) = Q_M^k(v).$$

Intuitively, M cannot distinguish k-equivalent suffixes when using k space. The following lemma makes it more formal.

Lemma 5.1.2. Let $x, u, v \in \Sigma^*$, and $u \equiv_k v$. Then the following two statements are equivalent:

- there is a k space-bounded accepting computation of M on xu
- there is a k space-bounded accepting computation of M on xv.

Proof. Let us consider some k space-bounded accepting computation of M on xu:

$$\beta_0 \beta_1 \ldots \beta_m,$$

where β_0 is the initial configuration and β_m is accepting. We shall show that there is also a k space-bounded accepting computation on xv. During the computation $\beta_0 \beta_1 \ldots \beta_m$ M can cross the boundary between x and u several times. So, the computation can be divided into segments: $\alpha_1 \alpha_2 \ldots \alpha_j$ in such a way that α_1 starts in β_0, α_j ends in β_m, and for each i, the segment α_i consists entirely of configurations entered on x if i is odd, and entirely of configurations entered on u if i is even. From the assumption that $u \equiv_k v$ it follows that for each segment α_i with i even, there exists a k space-bounded segment of computation δ_i on v that enters and exits v exactly in the same internal configurations as α_i; and if j is even, then there is also a k space-bounded accepting segment of computation δ_j that enters v exactly in the same internal configuration as α_j. Now, if we exchange each segment α_i with even i for the segment δ_i, we obtain an accepting computation of M on xv. $\Box$

Now we are ready to prove that $\log \log n$ is the lower bound for accepting non-regular languages by deterministic or nondeterministic Turing machines (Alberts 1985).

Theorem 5.1.3. Let M be a weakly $L(n)$ space-bounded deterministic or nondeterministic Turing machine. Then either

- $L(n) \geq c \log \log n$ for some constant $c > 0$ and infinitely many n, or
- M accepts a regular language and space used by M is bounded by a constant.

The theorem is also valid for the strong mode of space complexity.

Proof. Suppose that there is no constant upper bound on space used by M. Then there are infinitely many k and w such that $w = a_1 a_2 \ldots a_n$ is the shortest input accepted by M in exactly k space.

Define $w_i = a_i \ldots a_n$ to be a suffix of w, $1 \leq i \leq n$, and suppose that $w_i \equiv_k w_j$ and $w_i \equiv_{k-1} w_j$ for some $1 \leq i < j \leq n$. Then from Lemma 5.1.2 it follows that there is also k space-bounded accepting computation on the shorter word $w' = a_1 a_2 \ldots a_{i-1} a_j a_{j+1} \ldots a_n$. Note, that there is no accepting computation on w' that uses less space than k, since otherwise w could also be accepted with less space because $w_i \equiv_{k-1} w_j$.

By Lemma 2.3.2, the number of internal configurations using up to k space is at most d^k for some constant d. And the number of possible equivalence classes of the relation $\equiv_k$ is not greater than $4^{(d^k)^2}$. Since two different suffixes w_i and w_j cannot belong to the same equivalence classes of $\equiv_k$ and $\equiv_{k-1}$, it requires that

$$(4^{(d^k)^2})^2 \geq n$$

and taking logarithm twice we obtain $k \geq c \log\log n$ for some constant $c > 0$. Hence, $L(n) \geq c \log\log n$ for infinitely many n.   □

Now we shall present a modified version of the proof of Iwama (1993) that $\log\log n$ is the lower bound for accepting non-regular languages also for weakly space-bounded alternating Turing machines. Let M be an alternating Turing machine. We assume that all final configurations of M are encountered only on the right endmarker. Let an input w be of the form $w = xy$.

Let β be an internal configuration and i be a position on the input tape. Then an *accepting subtree* T rooted in β at i is a finite tree whose nodes are labeled by configurations of M on w. The root of the tree is labeled by β positioned at cell i, the children of any node labeled by a universal configuration are all immediate successors of that configuration; and any node labeled by an existential configuration has one child, which is labeled by one of the immediate successors of that configuration (provided there are any); and all leaves of the tree are accepting configurations.

$SEC_{x|y}(T)$ denotes the cross section of T which is the set of all internal configurations of T encountered just after M crosses the boundary between x and y for the first time. $CRS(T)$ denotes how many times computation paths of T cross the boundary. The number of crosses may differ from path to path of the tree. $CRS(T)$ is the maximum value over all paths. $SP(T)$ denotes the amount of space used by T.

We shall consider subtrees rooted on both sides of the boundary between x and y, i.e. either at the last letter of x or at the first letter of y. Note that, since all final configurations are entered on the right endmarker, $CRS(T)$ is odd if T is rooted at the last letter of x, and even if T is rooted at the first letter of y.

Now we define $S^k_{x|y}(r)$ for natural numbers k and r. Suppose first that r is even. Then $S^k_{x|y}(r)$ contains all pairs (β, j) of an internal configuration β and an integer $j \leq k$ which satisfy the following conditions:

- There exists an accepting subtree T rooted in β at the first letter of y such that $CRS(T) = r$ and $SP(T) = j$ (i.e. T crosses the boundary r times and uses j space).
- For any accepting subtree T' rooted in β at the first letter of y, if $CRS(T') = r$, then $SP(T') \geq j$.
- For any accepting subtree T' rooted in β at the first letter of y, if $CRS(T') < r$, then $SP(T') > j$.

Intuitively speaking, (β, j) is in $S^k_{x|y}(r)$, with even r, if β leads to acceptance with r crosses and j (but not less) space, and $(\beta, j') \notin S^k_{x|y}(r')$ for any $j' \leq j$ and even $r' < r$. Recall that $CRS(T)$ is even if T is rooted at the first letter of y.

If r is odd, then the definition is analogous, but subtrees are rooted at the last letter of x.

Lemma 5.1.4.

If r is even, then $S_{x|y}^k(r)$ is determined by y and by $S_{x|y}^k(1)$, $S_{x|y}^k(3)$, ... , $S_{x|y}^k(r-1)$. If r is odd, then $S_{x|y}^k(r)$ is determined by x and by $S_{x|y}^k(0)$, $S_{x|y}^k(2)$, ... , $S_{x|y}^k(r-1)$.

Proof. By the definition, $S_{x|y}^k(0)$ contains all (β, j) such that the internal configuration β positioned at the first letter of y leads to acceptance using j space and without crossing the boundary. Thus, clearly $S_{x|y}^k(0)$ depends only on y.

Now suppose that the lemma is valid for $S_{x|y}^k(0)$ through $S_{x|y}^k(r)$ for an even r. Let

$$MINS_{x|y}(r) = \{(\beta, j) \mid (\beta, j) \in S_{x|y}^k(0) \cup S_{x|y}^k(2) \cup \ldots \cup S_{x|y}^k(r) \quad \text{and}$$
$$(\beta, j') \notin S_{x|y}^k(0) \cup S_{x|y}^k(2) \cup \ldots \cup S_{x|y}^k(r) \text{ for any } j' < j\}.$$

Consider now $S_{x|y}^k(r+1)$. $(\beta, j) \in S_{x|y}^k(r+1)$ if the following conditions are satisfied:

(a) there exists an accepting subtree T rooted in β at the last letter of x satisfying the following conditions:
(i) $SEC_{x|y}(T) \subseteq \{\delta \mid \text{there is } m \text{ such that } (\delta, m) \in MINS_{x|y}(r)\}$ and
(ii) $j = \max\{m \mid \text{there is } \delta \text{ such that } \delta \in SEC_{x|y}(T) \text{ and}$
$(\delta, m) \in MINS_{x|y}(r)\}$
(b) there is no accepting subtree T' rooted in β at the last letter of x and $j' < j$ that satisfy (i) and (ii) with T' instead of T and j' instead of j,
(c) $(\beta, j') \notin S_{x|y}^k(r')$ for any $j' \leq j$ and odd $r' < r$ (by induction, this depends only on x and on $S_{x|y}^k(0)$, $S_{x|y}^k(2)$, ... , $S_{x|y}^k(r'-1)$).

Thus, the fact whether or not $(\beta, j) \in S_{x|y}^k(r+1)$ depends only on x and on $S_{x|y}^k(0)$, $S_{x|y}^k(2)$, ... , $S_{x|y}^k(r)$. For odd r, the proof is analogous.   □

Note that, if $r \neq r'$ and $r' + r$ is even (i.e. r and r' are both even or both odd), then $S_{x|y}^k(r) \cap S_{x|y}^k(r') = \emptyset$, and by Lemma 2.3.2, the number of k space-bounded internal configurations is bounded by d^k for some constant d. Hence, there may be at most $2kd^k$ nonempty $S_{x|y}^k(r)$. Thus, there exists an integer r such that $S_{x|y}^k(r) = \emptyset$. It is also obvious that if $S_{x|y}^k(r) = \emptyset$, then $S_{x|y}^k(r+1) = S_{x|y}^k(r+2) = \ldots = \emptyset$. Let $S_{x|y}^k(p)$ be the last one that is not empty. Then $S_{x|y}^k$ is defined as

$$S_{x|y}^k = (S_{x|y}^k(0),\ S_{x|y}^k(1),\ \ldots,\ S_{x|y}^k(p)).$$

Lemma 5.1.5. Let xyz be an input. Then $S_{x|yz}^k = S_{xy|z}^k$ implies $S_{x|yz}^k = S_{xy|z}^k = S_{x|z}^k$.

Proof. By Lemma 5.1.4, both $S_{xy|z}^k(0)$ and $S_{x|z}^k(0)$ depends only on z, therefore they must be the same. So by the assumption,

$$S^k_{xy|z}(0) = S^k_{x|yz}(0) = S^k_{x|z}(0).$$

This and again Lemma 5.1.4 imply that $S^k_{x|yz}(1) = S^k_{x|z}(1)$ and therefore

$$S^k_{xy|z}(1) = S^k_{x|yz}(1) = S^k_{x|z}(1).$$

We can continue this argument step by step until we get to

$$S^k_{xy|z}(p+1) = S^k_{x|yz}(p+1) = S^k_{x|z}(p+1) = \emptyset.$$

$\square$

Lemma 5.1.6. Let xyz be an input and $S^k_{x|yz} = S^k_{xy|z}$. Then

if M accepts xyz with k space but not with $k-1$ or less,
then M also accepts xz with k space and not with $k-1$ or less.

Proof. Let

$$MINS_{x|yz} = \{(\beta, j) \,|\, (\beta, j) \in S^k_{x|yz}(r) \text{ for some even } r, \text{ and}$$
$$(\beta, j') \notin S^k_{x;yz}(r) \text{ for any } j' < j \text{ and even } r \,\},$$

and suppose that M accepts xyz with k space and not with $k-1$.
  Then the following conditions must be satisfied:

(d) there is a computation tree T (rooted in the initial configuration) such that
  (iii) $SEC_{x|yz}(T) \subseteq \{\delta \mid \text{there is } m \text{ such that } (\delta, m) \in MINS_{x|yz}\}$ and
  (iv) $k = \max\{m \mid \text{there is } \delta \text{ such that } \delta \in SEC_{x|yz}(T) \text{ and}$
    $(\delta, m) \in MINS_{x|yz}\}$,
(e) there is no computation tree T' and $k' < k$ that satisfy (iii) and (iv) with
  T' instead of T and k' instead of k.

Let us recall that any final configurations can be entered by M only on the
right endmarker, so there are no rejecting leaves in computation trees T and T'
entered on x.
  Note that the sets

$$\{SEC_{x|yz}(T) \mid T \text{ is a computation tree on } xyz\} \quad \text{and}$$
$$\{SEC_{x|z}(T) \mid T \text{ is a computation tree on } xz\}$$

are equal and by Lemma 5.1.5, $S^k_{x|yz} = S^k_{x|z}$. Hence, conditions analogous to (d)
and (e) but with $x|z$ instead of $x|yz$ are satisfied on xz. Thus, M accepts xz
with space k and not with $k-1$. $\square$

Now we are ready to present the theorem of (Iwama 1993) which says that $\log\log n$ is the lower bound for accepting non-regular languages by alternating Turing machines.

Theorem 5.1.7. Let M be a weakly $L(n)$ space-bounded alternating Turing machine.
Then either

- $L(n) \geq c\log\log n$ for some constant $c > 0$ and infinitely many n, or
- M accepts a regular language and space used by M is bounded by a constant.

The theorem is also valid for the strong mode of space complexity.

Proof. Suppose that there is no constant upper bound on space used by M. Then there are infinitely many k and w such that w is the shortest input accepted by M in exactly k space. From Lemma 5.1.6 it follows that if $w = xyz$ and $y \neq \lambda$, then $S^k_{x|yz} \neq S^k_{xy|z}$, since otherwise we can get the shorter input xz that also needs k space.

By Lemma 2.3.2, the number of different k space-bounded internal configurations is bounded by d^k. Hence, there are at most 2^{kd^k} different $S^k_{x|y}(r)$ and

$$S^k_{x|y}(r) = \emptyset \qquad \text{if } r > 2kd^k.$$

Hence, the number of different $S^k_{x|y}$ is bounded by

$$(2^{kd^k})^{2kd^k} = 2^{2k^2d^{2k}}.$$

Thus

$$n \leq 2^{2k^2d^{2k}}.$$

So we can derive that $k \geq c\log\log n$ for some constant $c > 0$. Hence, $L(n) \geq c\log\log n$ for infinitely many n. $\square$

5.2. Lower Bounds for One-way Turing Machines

In this section we present lower bounds for accepting non-regular languages by one-way Turing machines. In this case the lower bound is equal to $\log n$ for strongly space-bounded deterministic, nondeterministic, and alternating, and weakly space-bounded deterministic Turing machines; and is equal to $\log\log n$ for weakly space-bounded nondeterministic and alternating Turing machines (Stearns et. al. 1965, Alberts 1985).

Theorem 5.2.1. Let M be a strongly $L(n)$ space-bounded one-way deterministic, nondeterministic or alternating Turing machine. Then either

- $L(n) \geq c\log n$ for some constant $c > 0$ and infinitely many n, or
- M accepts a regular language and space used by M is bounded by a constant.

Proof. Suppose that there is no constant upper bound on space used by M. Then there are infinitely many k and w such that

$$w = a_1 a_2 \ldots a_n$$

is the shortest input causing M to use exactly k space. Consider a computation that uses k space on w. The configuration using k space must not be reached before M scans a_n. Suppose that for some i and j, $1 \leq i < j \leq n$, there is an internal configuration which M encounters when shifting from a_i to a_{i+1} and from a_j to a_{j+1}. Then

$$a_1 a_2 \ldots a_i a_{j+1} \ldots a_n$$

is a shorter word causing M to use exactly k space.

By Lemma 2.3.2, the number of all k space-bounded internal configurations of M is bounded by d^k for some constant d. Hence,

$$n - 1 \leq d^k.$$

Taking logarithm we can derive that $k \geq c \log n$ for some constant $c > 0$. Hence, $L(n) \geq c \log n$ for infinitely many n. □

For weakly space-bounded one-way deterministic Turing machines the lower bound is $\log n$. The proof is essentially the same as for strongly space-bounded deterministic Turing machines in Theorem 5.2.1. One should only consider an accepting computation. For weakly space-bounded one-way nondeterministic and alternating Turing machines, the lower bound is $\log \log n$ because by Lemma 4.1.3, the non-regular language $\{a^k b^m \mid k \neq m\}$ can be accepted by a weakly $\log \log n$ space-bounded one-way nondeterministic Turing machine.

5.3. Lower Bounds for the Middle Mode of Space Complexity

In this section we consider lower bounds for the middle mode of space complexity, which requires that all computations on every accepted input, satisfy the space bounds. For two-way Turing machines, $\log \log n$ is the lower bound for accepting non-regular languages for both the weak and strong mode of space complexity. Hence, $\log \log n$ is also the lower bound for middle space-bounded deterministic, nondeterministic, and alternating two-way Turing machines.

The situation is more interesting for one-way Turing machines because then the lower bounds for the middle mode are different from both the lower bounds for the weak and strong mode. We shall show that the lower bound for the middle space-bounded one-way Turing machines is equal to $\log \log n$ for alternating, and to $\log n$ for deterministic and nondeterministic Turing machines (Szepietowski 1988).

Theorem 5.3.1. Let M be a middle $L(n)$ space-bounded one-way deterministic or nondeterministic Turing machine. Then either

- $L(n) \geq c \log n$ for some constant $c > 0$ and infinitely many n, or
- M accepts a regular language and space used by M is bounded by a constant.

Proof. Suppose that there is no constant upper bound on space used by M. Then there are infinitely many k and w such that

$$w = a_1 a_2 \ldots a_n$$

is the shortest accepted input causing M to use exactly k space. Consider two computations of M on w: one that uses exactly k space on w and the other that accepts w. Note that the accepting computation does not have to use maximal space. Suppose that both computations repeat internal configurations exactly in the same places. More precisely, for some i and j, $1 \leq i < j \leq n$, there are two internal configurations β_1 and β_2 such that:

- β_1 is entered by the accepting computation when shifting from a_i to a_{i+1} and from a_j to a_{j+1}, and
- β_2 is entered by the computation using k space when shifting from a_i to a_{i+1} and from a_j to a_{j+1}.

Then

$$a_1 a_2 \ldots a_i a_{j+1} \ldots a_n$$

is a shorter accepted word causing M to use exactly k space.

By Lemma 2.3.2, the number of all k space-bounded internal configurations of M is bounded by d^k for some constant d. We must therefore have

$$n - 1 \leq (d^k)^2.$$

Taking logarithm we can derive that $k \geq c \log n$ for some constant $c > 0$. Hence, $L(n) \geq c \log n$ for infinitely many n. □

For middle space-bounded one-way alternating Turing machines the lower bound is $\log \log n$ (Szepietowski 1988).

Theorem 5.3.2. Let

$$A = \{a^k b^m \mid m \text{ is a common multiple of all } r \leq k\}.$$

Then

- A is non-regular and
- $A \in middle\text{-}one\text{-}way\text{-}ASPACE[\log \log n]$.

Proof. First we show that A can be accepted by a middle $\log \log n$ space-bounded one-way alternating Turing machine M. M first writes in binary on the work tape the number k of a's standing at the beginning of the input w. Next, using universal branching selects a number $r \leq k$ and checks if r divides the number of b's. If $w = a^k b^m \in A$, then by Lemma 4.1.2, $m > c^k$ for some

$c > 1$, and M uses only $\log\log|w|$ space, though if $w \notin A$, it may happen that M uses too much space before it rejects w.

Now we shall show that A is non-regular. Assume that A can be accepted by a finite automaton with s states. Let k be such a number that $k > s$ and let m be a common multiple of all numbers $i \leq k$. Then $a^k b^m \in A$. By the pumping lemma 4.1.1, $a^k b^m$ can be decomposed into $a^k b^m = uvw$ in such a way that $|uv| \leq s$, $|v| \geq 1$, and $uv^{m+1} w \in A$. Note that $u, v \in a^*$, since $|uv| \leq s < k$. Hence, $uv^{m+1} w = a^{k+|v|m} b^m \in A$, a contradiction because $k + |v|m$ does not divide m. Thus A is non-regular. $\qquad\qquad\square$

6. Space Constructible Functions

In this chapter we consider space constructible and fully space constructible functions. Above $\log n$ all "normal" functions are fully space constructible. Below $\log n$, the situation is more complicated. Freedman and Ladner (1975) proved that neither $\log \log n$ nor any other unbounded non-decreasing $o(\log n)$ function is fully space constructible. Geffert (1991) proved the same for nondeterministically fully space constructible functions.

In Chapter 8 we shall show that there exist computable functions above logarithm which are not fully space constructible and in Chapter 15, we shall consider functions that are space constructed by other models of Turing machines.

6.1. Definitions and Basic Properties

Definition 6.1.1. *A function $L(n)$ on natural numbers is said to be space constructible if there exists a strongly $L(n)$ space-bounded deterministic Turing machine M such that for every n, there is an input of length n on which M uses exactly $L(n)$ space.*

We say that M constructs $L(n)$.

Definition 6.1.2. *$L(n)$ is called fully space constructible if there exists a deterministic Turing machine M such that on every input of length n, M uses exactly $L(n)$ space.*

We say that M fully constructs $L(n)$.

Definition 6.1.3. *A function $L(n)$ is nondeterministically fully space constructible if there exists a strongly $L(n)$ space-bounded nondeterministic Turing machine M such that on every input of length n, there is a computation of M which stops and uses exactly $L(n)$ space.*

Let us note that for $L(n)$ to be nondeterministically fully space constructible, it suffices that there exists a strongly $L(n)$ space-bounded nondeterministic Turing machine such that there is an accessible configuration using $L(n)$ space on every input of length n. In this case, one has only to introduce a new final state and the rule that this new state can follow any other state at any moment.

The following lemma presents the basic properties of space constructible and fully space constructible functions.

Lemma 6.1.4.

(a) If $L(n)$ is fully space constructible, then $L(n)$ is space constructible,

(b) if $L(n)$ is space constructible and $L(n) \geq n$, then $L(n)$ is fully space constructible,

(c) if $L(n)$ and $K(n)$ are (fully) space constructible, then $L(n) + K(n)$ and $L(n) \cdot K(n)$ are (fully) space constructible,

(d) if $L(n)$ and $K(n)$ are (fully) space constructible and $K(n) \geq n$, then the composition $K(L(n))$ is (fully) space constructible.

From Theorem 5.1.3 it follows that there are no $o(\log \log n)$ space constructible functions. But in some sense, the function $\log \log n$ is constructible.

Lemma 6.1.5.   There is a space constructible function $L(n)$ such that

$$\log \log n \leq L(n) \leq \log \log n + 1.$$

Proof.   Consider the deterministic Turing machine M, described in Chapter 4, which accepts the language consisting of all words of the form:

$$w_k = b_0 \# b_1 \# \ldots \# b_k,$$

where b_i is the binary description of the number i. M uses exactly $\lfloor \log \lfloor \log k \rfloor \rfloor + 1$ space if the input is of the form $w_k x$, with $x \in \{0, 1, \#\}^*$ which does not begin with $\# b_{k+1}$. Since $k < |w_k| < |w|$, the space used by M on $|w|$ does not exceed $\log \log |w| + 1$. On the other hand, for each n, there is an input w such that $|w| = n$, $w = w_k x$, and $|x| < |w_k|$. In this case

$$|w| < 2|w_k| < 2k \log k,$$

hence

$$\log \log |w| \leq \lfloor \log \lfloor \log k \rfloor \rfloor + 1$$

for sufficiently large k. For short inputs, M can construct the space function separately. $\square$

Note that by Lemma 6.1.4(d), any composition $K(\log \log n)$ is constructible if $K(n)$ is fully space constructible and $K(n) \geq n$. In particular, any function $\lfloor \log \log n \rfloor^r$ with rational $r \geq 1$ is space constructible.

6.2. Fully Space Constructible Functions

Above $\log n$, i.e. if $L(n) \geq c \log n$ for some constant $c > 0$, all "normal" functions are fully space constructible, e.g.: $\log n$ itself, n^r, $[\log n]^q$, and q^n with rational $r > 0$ and $q \geq 1$. Below $\log n$, the situation is more complicated. Freedman and Ladner (1975) proved the following theorem.

Theorem 6.2.1. If $L(n)$ is fully space constructible, then either

- $L(n) \geq c \log n$ for some constant $c > 0$ and all n, or
- $L(n) < c_1$ for another constant c_1 and infinitely many n.

Proof. Let M be a deterministic Turing machine that fully constructs $L(n)$ and suppose that $\liminf \frac{L(n)}{\log n} = 0$ or in other words that there is no constant $c > 0$ such that $L(n) \geq c \log n$ for all n. Let $r(n)$ denote the number of all possible $L(n)$ space-bounded internal configurations of M. By Lemma 2.3.2, $r(n) \leq d^{L(n)}$, for some constant d. Since $\liminf \frac{L(n)}{\log n} = 0$, there exists a number m such that $r(m) < m$. Consider the computation of M on a^m and on $a^{m+km!}$ for an arbitrary natural k. Whenever M traverses a^m from one end to the other, it has to loop (repeating internal configurations) and the length of the loop is less than m and divides $m!$. So, M reaches the other end in the same internal configuration on a^m and on $a^{m+km!}$. Similarly, if M starting from one endmarker of a^m comes back to the same endmarker without visiting the other endmarker, then M can do the same on $a^{m+km!}$. Thus, behavior of M on a^m and on $a^{m+km!}$ is essentially the same. It differs only in the number of loops performed during passes from one endmarker to the other. Hence, $L(m) = L(m + km!)$ for every k. Thus, there exists a constant $c_1 = L(m)$ such that $L(n) \leq c_1$ for infinitely many n. $\qquad\square$

The following corollary follows immediately from Theorem 5.1.3 and Theorem 6.2.1.

Corollary 6.2.2. If $L(n)$ is unbounded, fully space constructible, and $\liminf \frac{L(n)}{\log n} = 0$, then there exist two constants $d_1, d_2 > 0$ such that:

- $L(n) \geq d_1 \log \log n$ for infinitely many n, and
- $L(n) < d_2$ for infinitely many n.

Thus, neither $\log \log n$ nor any unbounded and non-decreasing $o(\log n)$ function is fully space constructible. But there exists unbounded $O(\log \log n)$ fully space constructible function.

Lemma 6.2.3. Let $L(n) = \log[\min\{i \mid i \text{ does not divide } n\}]$. Then $L(n)$ is fully space constructible and $L(n) = O(\log \log n)$.

Proof. First, we present a deterministic Turing machine M constructing $L(n)$. On an input of length n, M computes in binary $F(n) = \min\{i \mid i$ does not divide $n\}$. To do this, M writes in binary on its work tape one by one successive natural numbers up to the moment it finds one that does not divide n. In order to check if a number k divides n, M forms a counter modulo k on the work tape and going from one end of the input to the other counts the letters of the input. Thus, M uses exactly $L(n)$ space on every input of length n. And by Lemma 4.1.2(d), $F(n) \leq c \log n$, so $L(n) = O(\log \log n)$. $\qquad\qquad\square$

The method used in the proof of Theorem 6.2.1 is called $n \to n + n!$. We shall also use this method to prove the following theorem (Chang et. al. 1987).

Theorem 6.2.4. Let A be an infinite bounded language (i.e. an infinite subset of $a_1^* a_2^* \ldots a_k^*$) accepted by a weakly (or strongly) $L(n) = o(\log n)$ space-bounded nondeterministic (or deterministic) Turing machine. Then A contains an infinite regular subset.

Proof. We shall show the result for a language $A \subseteq a^* b^*$ only. The proof can be generalized in an obvious way. Let M be a weakly $L(n) = o(\log n)$ space-bounded nondeterministic Turing machine accepting A and let $r(n)$ denote the number of all possible $L(n)$ space-bounded internal configurations of M. By Lemma 2.3.2, there is a constant d such that $r(n) \leq d^{L(n)}$. Since $\lim \cdot \frac{L(n)}{\log n} = 0$ and A is infinite, there exists some $w = a^i b^j \in A$ of length $i + j = m$ such that $i > r(m)$ or $j > r(m)$. Suppose that $i > r(m)$ and consider an accepting computation of M on w which consists only of $L(m)$ space-bounded configurations. Similarly as in the proof of Theorem 6.2.1, one can show that there is also an accepting computation on $a^{km!} w = a^{km! + i} b^j$ for every k. Hence, A contains the infinite regular set $\{a^{km!} w \mid k \geq 1\}$. $\qquad\qquad\square$

Using the above theorem we show that the language $\{a^k b^k \mid k \geq 0\}$ cannot be accepted with $o(\log n)$ space.

Corollary 6.2.5. Let $B = \{a^k b^k \mid k \geq 0\}$ and $L(n) = o(\log n)$. Then $B \notin weak\text{-}NSPACE[L(n)]$.

Proof. The proof is obvious, since by Lemma 4.1.4, B does not contain any infinite regular subset. $\qquad\qquad\square$

Note that the language B can be accepted with space $L(n) \ll \log n$ (i.e. $\liminf \frac{L(n)}{\log n} = 0$). This is because every input with odd length is not in B and can be rejected without using any space. The following theorem shows that this is not possible if a language contains inputs of (almost) all lengths.

Theorem 6.2.6. Let A be a language such that:

- A is an infinite subset of $a_1^* a_2^* \ldots a_k^*$,

- A is accepted by a weakly $L(n)$ space-bounded nondeterministic Turing machine with $\liminf \frac{L(n)}{\log n} = 0$, and
- there is a word $w \in A$ of length $|w| = n$ for all but finitely many n.

Then A contains an infinite regular subset.

Proof. The proof is similar to that of Theorem 6.2.4. Since $\liminf \frac{L(n)}{\log n} = 0$, there exists m such that $r(m) < \frac{m}{2}$ and there is a word $w \in A$ of length m with the group of a's or b's longer than $r(m)$. $\qquad\qquad\square$

Corollary 6.2.7. Let $D = \{a^k b^m \mid m = k \text{ or } m = k-1\}$ and $\liminf \frac{L(n)}{\log n} = 0$. Then $D \notin weak\text{-}NSPACE[L(n)]$.

Proof. For every n, there is a word of length n in D, and similarly as in Lemma 4.1.4 one can show that D does not contain any infinite regular subset. $\qquad\square$

Now we show that the language $\{a^k b^k \mid k \geq 0\}$ can be accepted by weakly $\log \log n$ space-bounded alternating Turing machine.

Theorem 6.2.8.

$$\{a^k b^k \mid k \geq 0\} \in weak\text{-}ASPACE[\log \log n].$$

Proof. We shall demonstrate a weakly $\log \log n$ space-bounded alternating Turing machine M which accepts $\{a^k b^k \mid k \geq 0\}$. On an arbitrary input w, M first deterministically checks whether the input is of the form $w = a^k b^m$. Then it (existentially) guesses and writes in binary a number x and branching universally checks the following two facts:

(i) Whether $k = m(\bmod\ j)$ for every $j \leq d_1 x$, where d_1 is a proper constant (see below).
(ii) Whether $x \geq H(|w|)$, where $H(|w|) = \max\{F(s) \mid s \leq |w|\}$ and $F(k) = \min\{i \mid i \text{ does not divide } k\}$.

The latter can be done as follows. Branching universally, M moves to the ith cell of the input, for each $1 \leq i \leq |w|$. Next it guesses (existentially) $p \leq x$ and going from the i-th to first cell checks if p divides i, and accepts if this is not the case. Note that M has to store (in binary) p, but not i.

If $w = a^k b^k$, then the branch (i) accepts, and there exists $x = H(2k) = H(|w|)$ such that the branch (ii) also accepts. Thus, M accepts w in at most $\log \log |w|$ space, because x and j are stored in binary and by Lemma 4.1.2(g), $H(|w|) \leq c \log |w|$.

If $w = a^k b^m$ and $k \neq m$, then if $x < H(|w|)$, then the branch (ii) does not accept, and if $x \geq H(|w|)$ and $d_1 = \frac{c}{d}$, where c and d are the constants from Lemma 4.1.2(g), then, by Lemma 4.1.2(e), there exists r such that $k \neq m(\bmod\ r)$

and $r \leq c \log |w| = d_1 d \log |w| \leq d_1 H(|w|) \leq d_1 x$ and the branch (i) does not accept. Thus the input $w = a^k b^m$ is not accepted, if $k \neq m$. □

In the sequel, we shall use the following property of fully space constructible functions (Szepietowski 1989).

Theorem 6.2.9. Let $L(n)$ be an arbitrary unbounded fully space constructible function, $L(n) = o(\log n)$, $f(n)$ be a function, and $\lim f(n) = \infty$. Then the language $B(f) = \{a^n \mid L(n) \geq f(n)\}$ does not contain any infinite regular subset.

Proof. Suppose, for a contradiction, that there is an infinite regular subset $R \subseteq B(f)$. Then there exist two numbers p and q such that $a^{p+mq} \in R$ for all m.

Consider now the function

$$S(n) = \max_{n-q \leq y \leq n} \{L(y)\}.$$

It is easy to see that

- $S(n)$ is fully space constructible,
- $\lim S(n) = \infty$, since $\lim f(n) = \infty$ and $S(i) \geq L(p + mq) \geq f(p + mq)$ for every i such that $p + mq \leq i < p + (m+1)q$ (note that $a^{p+mq} \in B(f)$).
- $S(n) = o(\log n)$, since $\lim \frac{L(n)}{\log n} = 0$ and

$$\frac{S(n)}{\log n} = \max_{n-q \leq y \leq n} \left\{ \frac{L(y)}{\log n} \right\} \leq \max_{n-q \leq y \leq n} \left\{ \frac{L(y)}{\log y} \right\}.$$

But this contradicts Theorem 6.2.1. □

Note that if $f(n)$ is constant, then $B(f)$ is regular: modify some Turing machine that constructs $L(n)$ so that the computation is interrupted as soon as more than $f(n)$ cells are used. This yields a Turing machine with constant work space, i.e. a finite automaton, that accepts $B(f)$.

6.3. Nondeterministically Fully Space Constructible Functions

Nondeterministically fully space constructible functions were considered in (Szepietowski 1989) where it was proved that the following three conditions are equivalent:

- *weak-NSPACE*[$\log \log n$] = *strong-NSPACE*[$\log \log n$]
- $\{a^k b^m \mid k \neq m\} \in$ *strong-NSPACE*[$\log \log n$]
- $\log \log n$ is nondeterministically fully space constructible.

Geffert (1991) showed that these conditions are not valid because neither $\log \log n$ nor any other unbounded nondecreasing $o(\log n)$ function is nondeterministically fully space constructible. Now we present his proof.

Theorem 6.3.1. If $L(n)$ is nondeterministically fully space constructible, then either

- $L(n) \geq c \log n$ for some constant $c > 0$ and all n, or
- $L(n) < c_1$ for another constant c_1 and infinitely many n.

Proof. Let M be a strongly $L(n)$ space-bounded nondeterministic Turing machine that fully constructs $L(n)$ and suppose that $\liminf \frac{L(n)}{\log n} = 0$ or in other words that there is no constant $c > 0$ such that $L(n) \geq c \log n$ for all n. Let $r(n)$ denote the number of all possible $L(n)$ space-bounded internal configurations of M. By Lemma 2.3.2, there exists a constant d such that $r(n) \leq d^{L(n)}$. Let m be a number such that $m > [r(m)]^6$. Such a number exists since $\liminf \frac{L(n)}{\log n} = 0$.

Consider the behavior of the Turing machine M on a^m and on $a^{m+km!}$ for an arbitrary k. In Lemma 6.3.4 below we shall show that if a $L(m)$ space-bounded internal configuration β is accessible (from the initial configuration) on $a^{m+km!}$, then β is also accessible on a^m. Moreover, if β is accessible on the left or right endmarker on $a^{m+km!}$, then β is accessible on the left or right endmarker on a^m, respectively.

Using this lemma we can prove that M uses at most $L(m)$ space on every input of the form $a^{m+km!}$. Indeed, suppose that M reaches on $a^{m+km!}$ a configuration that is not $L(m)$ space-bounded and let β be the last $L(m)$ space-bounded internal configuration entered just before M overuses the space. Then by Lemma 6.3.4, β is also accessible on a^m and M overuses the space in the next step on this input. Thus, $L(m + km!) \leq L(m)$ for every k. □

In order to prove Lemma 6.3.4 we first consider the behavior of M on $a^{m+km!}$, for an arbitrary $k \geq 0$ (if $k = 0$ we consider the behavior on a^m), and prove two lemmas. Recall that our Turing machines cannot write blank symbol on the work tape and a $L(m)$ space-bounded internal configuration can only be reached by a $L(m)$ space-bounded computation.

Lemma 6.3.2. If there is a $L(m)$ space-bounded computation on $a^{m+km!}$ which starts in an internal configuration β_1 at a position p_1 and ends in an internal configuration β_2 at a position p_2, say $p_1 \leq p_2$, and which does not visit the endmarkers meanwhile, then there is a $L(m)$ space-bounded computation which starts in β_1 at p_1 and ends in β_2 at p_2 and which does not visit the cells more than $[r(m)]^2$ to the left of p_1 and more than $[r(m)]^2$ to the right from p_2.

Proof. We shall only show that there is a computation leading from β_1 at p_1 to β_2 at p_2 not visiting the cells to the right of $p_2 + [r(m)]^2$. Suppose that a computation visits some cells to the right of $p_2 + [r(m)]^2$. Let for each position

$p > p_2$, $\delta_1(p)$ denote a configuration encountered when M visits the position p for the first time during the computation, and let $\delta_2(p)$ denote a configuration encountered when M comes back to the position p. Then there are two positions p_3 and p_4, $p_2 < p_3 < p_4$, such that $\delta_1(p_3) = \delta_1(p_4)$ and $\delta_2(p_3) = \delta_2(p_4)$. In that case, we can obtain a shorter computation from β_1 at p_1 to β_2 at p_2 by cutting off two segments of the old computation:

- one from $\delta_1(p_3)$ at the position p_3 to $\delta_1(p_3)$ at the position p_4 and the other
- from $\delta_2(p_3)$ at the position p_4 to $\delta_2(p_3)$ at the position p_3.

By repeating this process we can obtain the computation leading from β_1 at p_1 to β_2 at p_2 not visiting the cells to the right of $p_2 + [r(m)]^2$.   □

Note that using Lemma 6.3.2 we can freely move segments of a computation, provided they are at the safe distance of at least $[r(m)]^2$ from the endmarkers. More precisely, if M can get from an internal configuration β_1 at a position p_1 to an internal configuration β_2 at a position p_1+p visiting neither of the endmarkers, then M can also get from β_1 at p_2 to β_2 at $p_2 + p$, provided $[r(m)]^2 < p_2$ and $p_2 + p \leq m + km! - [r(m)]^2$.

The following lemma is crucial for the whole theorem. Roughly speaking, it says that we only need to consider those computations of M that loop when they traverse through the whole input. This lemma has many useful consequences.

Lemma 6.3.3. If the input $a^{m+km!}$ can be traversed from one endmarker to the other by a $L(m)$ space-bounded computation beginning in an internal configuration β_1 and ending in β_2, then $a^{m+km!}$ can also be traversed by a computation beginning in β_1 and ending in β_2 such that from the position $[r(m)]^2$ to $m + km! - [r(m)]^4$ the computation consists of loops of length p for some $p < r(m)$.

Proof. Suppose that the input $a^{m+km!}$ can be traversed from left to right by a $L(m)$ space-bounded computation beginning in an internal configuration β_1 and ending in β_2. Since there are at most $r(m)$ $L(m)$ space-bounded internal configurations, the machine must repeat its internal configuration during the first $r(m)$ steps. Suppose that the computation consists of:

- the first segment s_1 not longer than $[r(m)]^2$,
- then some consecutive loops of length $p < r(m)$, and
- then the segment s_2 after the loops,

and that the computation contains the shortest segment s_2 satisfying the above conditions. In the computation segment s_2, no loop of any length q can occur more than p times. Suppose, for a contradiction, that there are at least p loops of length q. They are not necessarily adjacent, they can start in different internal configurations, but they all are of equal length q. But then by a "cut and paste" technique we can exchange p loops of length q for q loops of length p standing at the beginning. So, there is a computation from β_1 to β_2 with a shorter segment

s_2. These can be done because all cuttings and pastings are done at a safe distance of at least $[r(m)]^2$ from the endmarkers.

But there are no more than $[r(m)]^2$ loops with different length $q < r(m)$, each length can be repeated at most $r(m)$ times, and there is a loop in every segment of length $r(m)$. So, s_2 cannot be longer than $[r(m)]^4$.   □

Now we are ready to prove:

Lemma 6.3.4. If a $L(m)$ space-bounded internal configuration β is accessible (from the initial configuration) on $a^{m+km!}$, then β is also accessible on a^m.

Moreover, if β is accessible on the left or right endmarker on $a^{m+km!}$, then β is accessible on the left or right endmarker on a^m, respectively.

Proof. Suppose that there is a $L(m)$ space-bounded computation of M on $a^{m+km!}$ that leads from the initial configuration to an internal configuration β. We shall show that β is also accessible on a^m. The computation on $a^{m+km!}$ can be divided into segments of three types:

(a) segments that start and end at the same endmarker without visiting the other endmarker,

(b) segments that traverse the whole $a^{m+km!}$, and

(c) the final segment that reaches β without visiting any of the endmarkers.

By Lemma 6.3.2, each segment of type (a) can be exchanged for an equivalent segment that starts and ends in the same internal configurations and does not go further than $[r(m)]^2$ from the endmarker, the latter can also be performed on a^m. Suppose that M traverses $a^{m+km!}$ by a segment of type (b) that starts in an internal configuration β_1 at the beginning of the input and ends in an internal configuration β_2 at the end. Then by Lemma 6.3.3, it can be exchanged for a segment that starts in β_1 and ends in β_2 and that loops between $[r(m)]^2$ and $m + km! - [r(m)]^4$. Since the length of the loop, say p, divides $m!$, by cutting off $\frac{km!}{p}$ loops we obtain a computation on a^m that leads from β_1 at the beginning to β_2 at the end. This is possible, since $m > [r(m)]^6 > [r(m)]^4 + [r(m)]^2$. The segment of type (c) can be exchanged for the segment that reaches β not farther than $r(m)$ from the endmarker, and the same segment can be performed on a^m. Hence, we have proved that the internal configuration β can also be reached on a^m.   □

7. Halting Property and Closure under Complement

In this chapter we consider the halting property of Turing machines and closure under complement of space complexity classes. Using the well-know technique of (Immerman 1988; Szelepcsényi 1988; Sipser 1980) we shall show that every strongly $L(n)$ space-bounded Turing machine (deterministic with arbitrary $L(n)$, and nondeterministic with $L(n) \geq \log n$) can be simulated by a strongly $L(n)$ space-bounded recognizing Turing machine (deterministic or nondeterministic, respectively). A recognizing Turing machine has two special states s_a and s_r and reaches the state s_a on every accepted input, and the state s_r on every not accepted input, and on no input both s_a and s_r can be reached (see Chapter 2).

Using this fact we shall show that classes $strong\text{-}DSPACE[L(n)]$ for arbitrary $L(n)$, and $strong\text{-}NSPACE[L(n)]$ for $L(n) \geq \log n$, are closed under complement (Sipser 1980; Immerman 1988; Szelepcsényi 1988)

It is not known whether the class $strong\text{-}NSPACE[L(n)]$ is closed under complement, if $L(n) = o(\log n)$ and whether any strongly $L(n)$ space-bounded nondeterministic Turing machine M with $L(n) = o(\log n)$ can be simulated by a recognizing Turing machine. But due to the result of (Geffert 1991), such simulation is possible for arbitrary function $L(n)$, if the language acceptable by M is bounded, i.e., if $L(M) \subseteq a_1^* a_2^* \ldots a_k^*$. Thus, if a language of the form $A \subseteq a_1^* a_2^* \ldots a_k^*$ is accepted by a strongly $L(n)$ space-bounded nondeterministic Turing machine, then its complement A^c is also accepted by a strongly $L(n)$ space-bounded nondeterministic Turing machine. This means, in particular, that the class of tally languages accepted by strongly $L(n)$ space-bounded nondeterministic Turing machines is closed under complement.

On the other hand, weak complexity classes are generally not closed under complement. The class $weak\text{-}DSPACE[L(n)]$ for arbitrary $L(n)$, and $weak\text{-}NSPACE[L(n)]$ for $L(n) \geq \log n$, are closed under complement only if they are equal to the strong complexity classes. Furthermore, there exist deterministic and nondeterministic weak space complexity classes above logarithm which are not closed under complement; and all weak space complexity classes between $\log \log n$ and $\log n$ are not closed under complement.

7.1. Halting Property of Turing Machines with Logarithmic or Greater Space

Definition 7.1.1. *We say that a Turing machine is halting if for each input, each computation stops eventually.*

In other words, for each input, there is no infinite computations and the tree of accessible configurations is finite.

The halting property of strongly $L(n)$ space-bounded Turing machines with $L(n) \geq \log n$ was discussed by Hopcroft and Ullman (1969). They observed that if there is an accepting computation of the deterministic or nondeterministic machine, then there is also an accepting computation without repeated configurations. Similarly, if there is an accepting tree for an alternating Turing machine, then there is also an accepting tree such that no configuration is repeated on any computation path in the tree. Thus, if during some computation some configuration is repeated, then the computation can be stopped in a rejecting state,

By Lemma 2.2.2, the number of all possible configurations of an $L(n)$ space-bounded Turing machine M on an input of length n is not greater than

$$s(n + 2)L(n)t^{L(n)},$$

where s and t represent the number of states and the number of symbols of the work tape alphabet, respectively. And if any computation is longer than $s(n + 2)L(n)t^{L(n)}$, then some configuration has been repeated and the computation can be stopped in a rejecting state. In order to detect too long computations, the Turing machine M counts its steps in base d, where d is such that $d^{L(n)} \geq s(n + 2)L(n)t^{L(n)}$. Such a constant exists, because $L(n) \geq \log n$. The counter is reset to zero whenever M uses a new cell on the work tape. The counter requires at most $L(n)$ space. Hence, we have proved the following theorem.

Theorem 7.1.2. Let M be a strongly $L(n)$ space-bounded deterministic, nondeterministic, or alternating Turing machine, and $L(n) \geq \log n$. Then there is a halting strongly $L(n)$ space-bounded deterministic, nondeterministic, or alternating, respectively, Turing machine M^h that accepts the same language as M.

Observe that if M is deterministic then M^h is a recognizing Turing machine. It stops in an accepting state if the input is accepted and in a rejecting, if the input is not accepted. From this it follows immediately that the class strong-$DSPACE[L(n)]$ is closed under complement if $L(n) \geq \log n$. In the next section we shall show that this is true for arbitrary function $L(n)$.

7.2. Closure under Complement of Strong Deterministic Complexity Classes

The halting algorithm of (Hopcroft and Ullman 1967), described in the previous section, does not apply to Turing machines with sublogarithmic space because the number of all possible configurations $s(n + 2)L(n)t^{L(n)}$ cannot be remembered within $L(n)$ space if $\liminf \frac{L(n)}{\log n} = 0$. Sipser (1980) showed that for any strongly $L(n)$ space-bounded deterministic Turing machine, there is an equivalent halting (or recognizing) deterministic Turing machine. Now we present a modified version of the proof of Sipser.

Theorem 7.2.1. Let M be a strongly $L(n)$ space-bounded deterministic Turing machine. Then there exists a strongly $L(n)$ space-bounded recognizing deterministic Turing machine M_R which accepts the same language. M_R has two final states s_a, s_r and on arbitrary input w either

- M_R reaches the state s_a if and only if M accepts w, or
- M_R reaches the state s_r if and only if M does not accept w

Proof. The Turing machine M_R will check if M can reach an accepting configuration from the initial configuration. Without loss of generality we assume that there is only one accepting configuration in M (i.e., when accepting an input, M erases its work tape, returns the heads to their initial positions, and halts in the unique accepting state).

First we present a procedure for determining whether M reaches the accepting configuration by a computation which uses at most space k. Consider configurations of M using at most k space as nodes of a finite graph, each configuration in the graph is connected by an arc with its immediate successor. Because M is deterministic, the component of this graph which contains the accepting configuration is a finite tree rooted at the accepting configuration. M_R begins at the accepting configuration and performs a depth-first search of this tree, to determine if the initial configuration is a member. This search procedure can easily be implemented on a Turing machine M_R using k space. When M_R is visiting a node, its work tape contains a copy of M's work tape and its input head is in the same place as M's input head. Arcs of the tree are traversed forward by running M one step forwards. Arcs are traversed backward by running M one step backwards. At some point there may be several valid ways to run M backwards. These are ordered in some predetermined way and M_R selects them one at a time in the usual depth-first manner. If, however, there are no valid ways to run M backwards, either because the current configuration has no predecessor or because the predecessors use $k + 1$ space, then a leaf of the tree has been reached and the search proceeds forwards.

A difficulty arises in utilizing the above procedure because M_R has no advance knowledge of k, the amount of space used by M. This is overcome by running the procedure successively for $k = 1, 2, 3, \ldots$ until an accepting computation is found. M_R uses exactly as much space as M does, provided M accepts. However, if M does not accept, then M_R will run forever. So, instead, M_R ensures that M uses at least $k + 1$ space before embarking on a $k + 1$ search. It does so by cycling through all configurations using space k and selecting those which are about to use the $(k + 1)$-th cell of the work tape in the next step. From each of these configurations, it performs a reverse search for the initial configuration. If none is successful, then M_R rejects, since it knows that M does not use space $k + 1$. On the other hand, if one of these searches succeeds, then M_R performs a $k + 1$ search from the accepting configuration, as before.

Now we shall explain how M_R cycles through all configuration of M which are about to use $k + 1$ space. M_R cycles through all k space-bounded internal configurations and positions of the input tape. For each k space-bounded internal configuration β, M_R cycles through all input head positions p starting from $p = 1$. At the beginning M_R stays at p in β, M_R keeps in its memory β and the symbol σ written at the position p (the position p is not remembered). If M in β at σ is about to use $k + 1$ space in the next step, then M_R performs a reverse search for the initial configuration. The search does not enter any $k + 1$ space-bounded configuration and ends when M_R comes back to the internal configuration β standing at the symbol σ. During the reverse search M_R enters β at σ only when it comes back to the position p. Then M_R proceeds search for the position $p + 1$.   □

From the above theorem it follows immediately that the strong deterministic complexity classes are closed under complement.

Theorem 7.2.2.   For any function $L(n)$,

$$strong\text{-}DSPACE[L(n)] = co\text{-}strong\text{-}DSPACE[L(n)].$$

Observe that the Turing machine M_R can be used (after small modification) to check whether the machine M accepts an input within given space. We shall need this fact in the sequel and we present it now more formally.

Theorem 7.2.3.   For every deterministic Turing machine M there exists a deterministic Turing machine $R(M)$ with two final states s_a, s_r such that, if $R(M)$ is started on arbitrary input w with arbitrary number s of cells marked off at the beginning of the work tape, then either

- $R(M)$ reaches the state s_a if and only if M accepts w within space s (if there is an s space-bounded accepting computation of M on w), or
- $R(M)$ reaches the state s_r if and only if M does not accept w within space s (w is not accepted at all or M needs more than s space to accept w).

During its computation $R(M)$ does not use more than s space.

Sipser's algorithm shows how to check whether there is an accepting computation starting from the initial configuration, but it cannot be used to find out whether there is an accepting computation starting from an arbitrary configuration. Now we shall present the algorithm of (Hartmanis and Berman 1976) which can be applied to computations starting in any configurations, but it is restricted to bounded languages (i.e. to languages of the form $A \subseteq a_1^* a_2^* \ldots a_k^*$).

Theorem 7.2.4. If a language $A \subseteq a_1^* a_2^* \ldots a_k^*$ is accepted by a strongly $L(n)$ space-bounded deterministic Turing machine M, then A^c –the complement of A, is also accepted by some strongly $L(n)$ space-bounded deterministic Turing machine M^c.

Moreover, for every $L(n)$ space-bounded configuration β of M, M starting from β accepts if and only if M^c starting from β does not accept (the work tape alphabet of M^c contains the work tape alphabet of M and the set of states of M^c contains the states of M).

Proof. We shall prove the theorem for $A \subseteq a^* b^*$. The proof can be generalized in an obvious way.

From Lemma 2.3.2 it follows that the Turing machine M cannot enter more than $d^{L(n)}$ internal configurations for some constant d. The Turing machine M^c accepting A^c acts as follows: M^c has six tracks on its work tape such that on each track, M^c using m space can count higher than d^m. On the track 1, M^c simulates M and if M ever halts, M^c rejects the input if M accepts, and vice versa, M^c accepts if M rejects. On the tracks 2, 3, and 4, M^c counts the number of times the input head of M hits the left and right endmarker, and the boundary between a's and b's, respectively. If either of these counters grows so large that it tries to use more space than so far used by M, M^c accepts the input, since M is cycling. On the track 5, M^c counts the number of moves M has performed since its input head last visited an endmarker or the boundary between a's and b's. If this counter tries to use more space than M has used so far, then M has repeated its internal configuration inside a's (or b's) and either is heading for another end of a's (or b's) or cycling near one end of a's (or b's). M^c now records the configuration of M on the track 6 and counts on the track 5 the displacement of the input head of M from its present position until the recorded configuration of M is repeated (which must happen in less steps than we can count on the available tape). If the displacement is zero, then M is cycling and M^c accepts the input; if the displacement is not zero, then the input head of M will eventually hit another end of a's (or b's) and the process starts all over. Since M uses $L(n)$ space on every input w of length n, M^c will eventually halt, accept w if and only if M does not, and use no more space than M. Thus, M^c accepts A^c strongly in $L(n)$ space.

Additionally if M^c is started in an arbitrary $L(n)$ space-bounded configuration β of M and β is positioned somewhere in the middle of a's or b's, then M^c has to count the number of moves M has performed since the start and the displacement of the input head from the starting position in order to detect if β is not in the middle of some loop. $\Box$

7.3. Closure under Complement of Strong Nondeterministic Space Complexity Classes above Logarithm

Using the proof technique of (Immerman 1988; Szelepcsényi 1988) one can show that also any $L(n)$ space-bounded nondeterministic Turing machine with $L(n) \geq \log n$ can be simulated by a recognizing nondeterministic Turing machine.

Theorem 7.3.1. Let M be a strongly $L(n)$ space-bounded nondeterministic Turing machine, and $L(n) \geq \log n$. Then there exists a strongly $L(n)$ space-bounded recognizing nondeterministic Turing machine M_R which accepts the same language. M_R has two final states s_a, s_r and on arbitrary input w either

- M_R reaches the state s_a if and only if M accepts w or
- M_R reaches the state s_r if and only if M does not accept w

Proof. First, we assume that $L(n)$ is fully space constructible. Let w be an input of length n and N be the exact number of $L(n)$ space-bounded configurations reachable by M from the initial configuration. The Turing machine M_R computes the number N. Let N_d be the number of configurations reachable in at most d steps. The computation proceeds by calculating N_0, N_1, and so on. If M_R has already computed N_d, it initializes a counter to 0 and cycles through all, so-called, target $L(n)$ space-bounded configurations in lexicographical order (M_R cycles through all words of length at most $L(n)$ on the work tape of M, all states, and all positions of the input and work heads). For each target configuration, M_R checks if it can be reached in less than d steps. To do this, M_R cycles through all N_d configurations (it cycles again through all configurations in lexicographical order, guesses which of them are in N_d, guesses a computation of length at most d for each reachable one, counts those reachable in at most d steps, and if it does not find all N_d of them, it rejects). If M_R finds the target configuration among these N_d or those which are reachable in one step more, then it increases the counter by one. When M_R completes this algorithm for all targets the counter contains N_{d+1}. By Lemma 2.2.2, N is bounded by $c^{L(n)}$ for some constant $c > 1$, hence, the space needed is bounded by $c_1 L(n)$ for some constant c_1 and c_1 depends only on the number of states and the number of work tape symbols of M. The exact number N of reachable configurations is equal to the first N_d such that $N_d = N_{d+1}$. Note that M_R while computing N goes through all reachable configurations of M and if it finds an accepting one, then it stops in the state s_a, and if not, in s_r. M_R reaches the state s_a if and only if M accepts, and the state s_r if and only if M does not accept.

    If $L(n)$ is not fully space constructible, M_R first assumes $L(n) = \log n$ (recall that $\log n$ is fully space constructible). If M accepts within the space $L(n) = \log n$, then M_R accepts. If M does not accept within the space $\log n$ and if it does not reach any configuration which is about to use the $\log n + 1$ cell of the work tape in the next step, then M_R rejects. If M reaches a configuration which is about to use the $\log n+1$ cell in the next step, then M_R assumes $L(n) = \log n+1$, and so on. If M is strongly $L(n)$ space-bounded, then M_R will eventually find the correct value of $L(n)$.   □

Using this theorem one can prove that strong nondeterministic complexity classes above logarithm are closed under complement (Immerman 1988; Szelépcsenyi 1988).

Theorem 7.3.2. For any function $L(n) \geq \log n$,

$$strong\text{-}NSPACE[L(n)] = co\text{-}strong\text{-}NSPACE[L(n)].$$

Proof. Let A be a language accepted by a strongly $L(n)$ space-bounded nondeterministic Turing machine M and $L(n) \geq \log n$. By Theorem 7.3.1, A is also accepted by an $L(n)$ space-bounded nondeterministic recognizing Turing machine M_R. If we exchange in M_R the accepting state s_a for rejecting s_r and vice versa, then the new machine will accept the complement of A.   □

Similarly as for deterministic Turing machines, M_R can be used to check whether the machine M accepts an input within given space.

Theorem 7.3.3. For every nondeterministic Turing machine M there exists a nondeterministic Turing machine $R(M)$ with two final states s_a, s_r such that, if $R(M)$ is started on arbitrary input w with arbitrary number $s \geq \log|w|$ of cells marked off at the beginning of the work tape, then either

- $R(M)$ reaches the state s_a if and only if M accepts w within space s (if there is an s space-bounded accepting computation of M on w), or
- $R(M)$ reaches the state s_r if and only if M does not accept w within space s (w is not accepted at all or M needs more than s space to accept w).

During its computation $R(M)$ does not use more than s space.

7.4. Closure under Complement for Bounded Languages Acceptable by Nondeterministic Turing Machines

It is not known whether the class $strong\text{-}NSPACE[L(n)]$ is closed under complement, if $L(n) = o(\log n)$, and whether any strongly $L(n)$ space-bounded nondeterministic Turing machine M can be simulated by a recognizing Turing machine. It is only known that such simulation is possible if the language acceptable by M is bounded, i.e., if $L(M) \subseteq a_1^* a_2^* \ldots a_k^*$ (Geffert 1991).

Theorem 7.4.1. Let M be a strongly $L(n)$ space-bounded nondeterministic Turing machine such that $L(M) \subseteq a_1^* a_2^* \ldots a_k^*$. Then there exists a strongly $L(n)$

space-bounded recognizing nondeterministic Turing machine M_R which accepts the same language. M_R has two final states s_a, s_r and on arbitrary input w either

- M_R reaches the state s_a if and only if M accepts w or
- M_R reaches the state s_r if and only if M does not accept w

Proof. Without loss of generality we can assume that $L(M) \subseteq a^*b^*$. Let w be an input of the form $w = a^k b^m$ and of length $n = k + m$. Similarly as in the proof of Theorem 7.3.1, M_R counts accessible configurations of M, but here M_R counts only the configurations with positions of the input head in the *inshore* sectors, i.e. at a distance of at most $[r(n)]^4$ from either of the endmarkers or from the boundary between a's and b's, where $r(n)$ denotes the number of all possible $L(n)$ space-bounded internal configurations of M. By Lemma 2.3.2, $r(n) \leq d_1^{L(n)}$ for some constant $d_1 > 1$.

Note that the number of all configurations with the input head in the inshore sectors is not greater than $d^{L(n)}$ for some d, and the numbers up to $d^{L(n)}$ can be kept in the memory using $L(n)$ space. Similarly, M_R can keep in the memory any configuration from the inshore sectors. In this case M_R remembers only the distance between the input head and the nearest endmarker or boundary.

We modify the notion of immediate successor of a configuration staying in an inshore sector. We say that a configuration β' is an immediate successor of another configuration β if

(a) β' can be reached in one step from β or
(b) β has its input head position at the distance of $[r(n)]^4$ from one end of a's (or b's); β' has its input head position at the distance of $[r(n)]^4$ from the other end of a's (or b's, respectively) and β' can be reached from β without going outside a's (or b's).

Now we shall show how M_R recognizes whether two configurations β and β' satisfy (b). Suppose that the input head position of β is $[r(n)]^4$ and the input head position of β' is $k - [r(n)]^4$ (on the other end of a's). Other cases can be worked out similarly. From the proof of Lemma 6.3.3 it follows that it suffices to consider only a configuration that can be reached by looping (i.e. by repeating internal configurations) because if there is an accepting computation of M on w, then there is also an accepting computation that always loops through the *deep sea* sectors of a's and b's, i.e. through a's and b's that are further than $[r(n)]^4$ from the endmarkers or the boundary between a's and b's. In order to check if β' can be reached from β by looping, M_R checks one by one for each pair of numbers i and j, $1 \leq i < j \leq r(n)$, if there are two configurations β_1 and β_2 such that

- the input head position of β_1 is equal to $[r(n)]^4 + i$,
- the input head position of β_2 is equal to $[r(n)]^4 + j$,
- both the internal configurations of β_1 and β_2 are equal to the internal configuration of β',

- β_1 can be reached from β in less than $r(n)$ steps,
- β_2 can be reached from β_1 in less than $r(n)$ steps.

If there are two such configurations, M_R checks if the distance between $[r(n)]^4$ and $k-[r(n)]^4$ is equal to $i+q(j-i)$ for some natural q. In order to check if β_1 is accessible from β (or if β_2 is accessible from β_1) in less than $r(n)$ steps, M_R uses the Immerman Szelepcsényi technique of counting all accessible configurations that can be reached in less than $r(n)$ steps (see Theorem 7.3.1).   □

Using the above theorem one can easily show the following theorem which was announced (without any proof) by Geffert (1991).

Theorem 7.4.2. Let $A \subseteq a_1^* a_2^* \ldots a_k^*$ and $L(n)$ be an arbitrary function, then

if $A \in strong\text{-}NSPACE[L(n)]$ then $A^c \in strong\text{-}NSPACE[L(n)]$,

where A^c denotes the complement of A.

Similarly as for nondeterministic Turing machines with logarithmic or greater space, M_R can be used to check whether the machine M accepts an input within given space.

Theorem 7.4.3. For any nondeterministic Turing machine M accepting a bounded language (i.e. if $L(M) \subset a_1^* a_2^* \ldots a_k^*$) there exists a nondeterministic Turing machine $R(M)$ with two final states s_a, s_r such that, if $R(M)$ is started on arbitrary input w with arbitrary number s of cells marked off at the beginning of the work tape, then either

- $R(M)$ reaches the state s_a if and only if M accepts w within space s (if there is an s space-bounded accepting computation of M on w), or
- $R(M)$ reaches the state s_r if and only if M does not accept w within space s (w is not accepted at all or M needs more than s space to accept w).

During its computation $R(M)$ does not use more than s space.

7.5. Nonclosure under Complement of Language Acceptable by Weakly Space-bounded Turing Machines

In this section we show that deterministic weak space complexity classes and nondeterministic weak space complexity classes above logarithm are closed under complement if and only if they are equal to the strong space complexity classes. Furthermore, we show that there exist deterministic and nondeterministic weak space complexity classes above logarithm which are not closed under complement, and that all weak space complexity classes between $\log \log n$ and

$\log n$ are not closed under complement. First we show that any weak nondeterministic space complexity class above logarithm is closed under complement if and only if it is equal to the strong space complexity class.

Theorem 7.5.1. For any function $L(n) \geq \log n$, the class $weak\text{-}NSPACE(L(n))$ is closed under complement if and only if

$$weak\text{-}NSPACE(L(n)) = strong\text{-}NSPACE(L(n)).$$

Proof. The "if" part of the theorem is obvious, since by Theorem 7.3.2 the class $strong\text{-}NSPACE(L(n))$ is closed under complement.

To prove the "only if" part suppose that a language $A \in weak\text{-}NSPACE$ $(L(n))$. Let M be a weakly $L(n)$ space-bounded nondeterministic Turing machine accepting A, and let M_c be a weakly $L(n)$ space-bounded nondeterministic Turing machine accepting A^c, the complement of A (by the assumption $A^c \in weak\text{-}NSPACE(L(n))$).

Now we construct a strongly $L(n)$ space-bounded Turing machine M_s which accepts A. On any input w the machine M_s first marks off $s = \log(|w|)$ cells at the work tape (this is possible, since the function $\log n$ is fully space constructible). Then M_s checks whether the machine M or M_c accepts w within space s. If one of them accepts, then M_s stops, in an accepting state, if w is accepted by M, or in a rejecting state, otherwise. If neither of the machines accepts, then M_s marks off one cell more and repeats the process.

More precisely, first M_s simulates on w the Turing machine $R(M)$, which is described in Theorem 7.3.3. If $R(M)$ reaches the state s_a (this means that M accepts w within space s), then M_s accepts the input w. If $R(M)$ reaches the state s_r (M does not accept w within space s), then M_s simulates the machine $R(M_c)$. If $R(M_c)$ reaches the state s_a, then M_s rejects the input w; if $R(M_c)$ reaches the state s_r, then M_s marks off one cell more and repeats the process.

The Turing machine M_s accepts the language A. It accepts an input w if and only if it finds out that w is accepted by M. On the other hand if $w \in A$, then it is accepted by M and not by M_c. Hence, M_s finds eventually that $w \in A$ and accepts it.

Furthermore M_s is strongly $L(n)$ space-bounded, because for every input w there is an accepting $L(|w|)$ space-bounded computation of M or M_c, and the machine M_s marks off a new $(s+1)$-th cell on the work tape only if it finds out that neither M nor M_c accepts w within space s. Remember that during simulation of $R(M)$ or $R(M_c)$ the machine M_s does not use more than the marked space. $\square$

In a similar way one can show that any weak deterministic space complexity class is closed under complement if and only if it is equal to the strong space complexity class.

Theorem 7.5.2. For any function $L(n)$, the class $weak\text{-}DSPACE(L(n))$ is closed under complement if and only if

$$weak\text{-}DSPACE(L(n)) = strong\text{-}DSPACE(L(n)).$$

Proof. The proof is exactly the same as the proof of Theorem 7.5.1 with two exceptions: Here we use Theorem 7.2.3 instead of Theorem 7.3.3, and the Turing machine M_s starts with $s = 1$ cells marked off at the work tape.   □

Now we show that there exist weak deterministic space complexity classes which are not closed under complement.

Theorem 7.5.3. Let $L_1(n)$ and $L_2(n)$ be two fully space constructible functions, $L_1(n) \leq L_2(n)$, and $\lim \frac{L_1(n)}{L_2(n)} = 0$. Then there exists a function $L_D(n)$ such that:

(1) $L_1(n) \leq L_D(n) \leq L_2(n)$.
(2) The function $L_D(n)$ is computable by an $L_2(n)$ space-bounded deterministic Turing machine M_D (in a sense that for every input of length n, M_D stops with $L_D(n)$ cells marked off at the beginning of the work tape).
(3) $weak\text{-}DSPACE(L_D(n))$ is not closed under complement.

Proof. By Theorem 11.1.1, there is a language over one-letter alphabet, $A \subset \{1\}^*$, which is accepted by some (strongly) $L_2(n)$ space-bounded deterministic Turing machine M_s and not accepted by any (strongly) $L_1(n)$ space-bounded one. And let

$$L_D(n) = \begin{cases} L_2(n) & \text{if } 1^n \in A \\ L_1(n) & \text{otherwise} \end{cases}$$

(1) follows immediately from the definition and from the fact that $L_1(n) \leq L_2(n)$.

(2) The Turing machine M_D which computes the function $L_D(n)$ works as follows. On any input w of the length n, M_D first marks off $L_2(n)$ cells at the work tape. Then it simulates the machine $R(M_s)$ working on 1^n (The machine $R(M_s)$ is described in Theorem 7.2.3 and M_s is the $L_2(n)$ space-bounded deterministic Turing machine which accepts A). During simulation all symbols on the input tape are read as 1. If $R(M_s)$ reaches the state s_a (this means that 1^n is accepted by M_s and $1^n \in A$), then M_D marks off $L_2(n)$ cells at the beginning of the work tape. If $R(M_s)$ reaches the state s_r (M_s does not accept 1^n), then M_D clears up the work tape and marks off $L_1(n)$ cells (both $L_1(n)$ and $L_2(n)$ are fully space constructible). Note that if $1^n \notin A$, then M_s uses more space than it is marked off at the work tape at the end.

To prove (3) observe that

$$A \in weak\text{-}DSPACE(L_D(n)).$$

Indeed, the $L_2(n)$ space-bounded Turing machine M_s which accepts A is also weakly $L_D(n)$ space-bounded, because on every accepted input $1^n \in A$ it uses at most $L_2(n)$ space and $L_2(n) = L_D(n)$ then. On the other hand, the complement of A:

$$A^c \notin weak\text{-}DSPACE(L_D(n)),$$

for otherwise there would also exist a strongly $L_1(n)$ space-bounded deterministic Turing machine M_1 which accepts the language A. Suppose, for a contradiction, that there exists a weakly $L_D(n)$ space-bounded deterministic Turing machine M_c which accepts the language A^c. Then there also exists a strongly $L_1(n)$ space-bounded deterministic Turing machine M_1 which accepts the language A. On every input 1^n, M_1 first marks off $L_1(n)$ cells at the beginning of the work tape ($L_1(n)$ is fully space constructible). Then it simulates on 1^n the Turing machine $R(M_c)$ (which is described in Theorem 7.2.3). M_1 accepts, if $R(M_c)$ reaches the state s_r (this means that M_c does not accept 1^n within $L_1(n)$ space). M_1 is strongly $L_1(n)$ space-bounded and accepts A. Indeed, if $1^n \in A$, then $1^n \notin A^c$, and 1^n is not accepted by M_c, and is accepted by M_1. If $1^n \notin A$, then 1^n is accepted by M_c within $L_D(n)$ space and $L_D(n) = L_1(n)$ then, so 1^n is not accepted by M_1. □

Now we show that there exist weak nondeterministic space complexity classes which are not closed under complement.

Theorem 7.5.4. Let $L_1(n)$ and $L_2(n)$ be two fully space constructible functions, $L_1(n) \leq L_2(n)$, and $\lim \frac{L_1(n)}{L_2(n)} = 0$. Then there exists a function $L_N(n)$ such that:

(1) $L_1(n) \leq L_N(n) \leq L_2(n)$.
(2) The function $L_N(n)$ is computable by an $L_2(n)$ space-bounded nondeterministic Turing machine M_N (in a sense that, for every input w of length n, M_N reaches an accepting configuration and all accepting configurations on w, have $S_N(n)$ cells marked off at the beginning of the work tape).
(3) $weak\text{-}NSPACE(L_N(n))$ is not closed under complement.

Proof. The proof of this theorem is similar to the proof of Theorem 7.5.3. Due to Theorem 11.1.1, there exists a language A over one-letter alphabet that is accepted by some (strongly) $L_2(n)$ space-bounded nondeterministic Turing machine and not accepted by any (strongly) $L_1(n)$ space-bounded one. Let

$$S_N(n) = \begin{cases} L_2(n) & \text{if } 1^n \in A \\ L_1(n) & \text{otherwise} \end{cases}$$

The rest of the proof is essentially the same as the proof of Theorem 7.5.3, we only use Theorem 7.4.3 instead of Theorem 7.2.3. □

Now we shall show that weak space complexity classes between $\log\log n$ and $\log n$ both deterministic and nondeterministic are not closed under complement.

Theorem 7.5.5. If $L(n) \geq \log\log n$ and $L(n) = o(\log n)$, then neither $weak\text{-}DSPACE[L(n)]$ nor $weak\text{-}NSPACE[L(n)]$ is closed under complement.

Proof. Consider the language

$$A = \{a^k b^m \mid k \neq m\}.$$

By Lemma 4.1.3,

$$A \in weak\text{-}DSPACE[\log \log n].$$

On the other hand, the complement of A

$$A^c \notin weak\text{-}NSPACE[L(n)]$$

because the class $weak\text{-}NSPACE[L(n)]$ is closed under intersection with regular languages, and by Corollary 6.2.5, the language

$$\{a^k b^k \mid k \geq 0\} = A^c \cap a^* b^* \notin weak\text{-}NSPACE[L(n)].$$

$\square$

Now we shall show that Theorem 7.5.5 is also valid if $L(n) \ll \log n$.

Theorem 7.5.6. If $L(n) \geq \log \log n$ and $\liminf \frac{L(n)}{\log n} = 0$, then neither $weak\text{-}DSPACE[L(n)]$ nor $weak\text{-}NSPACE[L(n)]$ is closed under complement.

Proof. Consider the language

$$C = \{a^k b^m \mid m \neq k \text{ and } m \neq k - 1\}$$

Using Corollary 6.2.7, similarly to the proof of Theorem 7.5.5, one can prove that

$$C \in weak\text{-}DSPACE[\log \log n]$$

and the complement of C

$$C^c \notin weak\text{-}NSPACE[L(n)].$$

$\square$

8. Strong Versus Weak Mode of Space Complexity

In this chapter we consider the relations between the weak and strong modes of space complexity. These modes are equivalent if the space bound function is fully space constructible. Thus, they are equivalent for all "normal" functions above logarithm. But there exist weak deterministic and nondeterministic space complexity classes, even above logarithm, which are not equal to the strong ones, and there exist computable functions above logarithm which are not fully space constructible. Finally, deterministic, nondeterministic, and alternating weak $L(n)$ space complexity classes are not equal to the strong ones, if $L(n) \geq \log \log n$ and $L(n) = o(\log n)$.

In Section 8.4 we shall also show that to prove that a recognizing nondeterministic Turing machine M is strongly $L(n)$ space-bounded, it suffice to show that M has at least one accepting or rejecting $L(n)$ space-bounded computation on every input. And in Section 8.5 we shall show that weak and middle modes of space complexity are equivalent for nondeterministic Turing machines with logarithmic or greater space.

8.1. Weak and Strong Mode of Space Complexity for Fully Space Constructible Functions

Theorem 8.1.1. Let $L(n)$ be fully space constructible, then

$$weak\text{-}XSPACE[L(n)] = strong\text{-}XSPACE[L(n)],$$

where X stands for D, N, or A.

Proof. Let M_c be a deterministic Turing machine that fully constructs $L(n)$ and M be a weakly $L(n)$ space-bounded deterministic, nondeterministic, or alternating Turing machine. We shall show that there is also a strongly $L(n)$ space-bounded deterministic, nondeterministic, or alternating, respectively, Turing machine M' which accepts the same inputs as M does. First, M' simulating M_c on w marks off $L(|w|)$ cells on its work tape. Then M' simulates M within space $L(|w|)$ and accepts if M accepts. Whenever M tries to use more than $L(|w|)$ space, M' stops in a rejecting state. It is obvious that M' accepts only those inputs that are accepted by M. On the other hand, if w is accepted by M, then there is an $L(|w|)$ space-bounded accepting computation (or tree for alternating Turing machine) on w and w is accepted by M'. Thus, M' accepts exactly the same inputs as M. $\square$

8.2. Weak and Strong Complexity Classes above Logarithm

Remember that above $\log n$ all "normal" functions are fully space constructible, e.g.: $\log n$ itself, n^r, $[\log n]^q$, and q^n with rational $r > 0$ and $q \geq 1$. But there are also weak deterministic and nondeterministic complexity classes above logarithm which are not equal to the strong ones, and there are computable functions above logarithm which are not fully space constructible.

Theorem 8.2.1. Let $L_1(n)$ and $L_2(n)$ be two fully space constructible functions, $L_1(n) \leq L_2(n)$, and $\lim \frac{L_1(n)}{L_2(n)} = 0$. Then there exist computable functions $L_D(n)$ and $L_N(n)$ such that:

(1) $L_1(n) \leq L_D(n)$, $L_N(n) \leq L_2(n)$.
(2) $weak\text{-}DSPACE(L_D(n)) \neq strong\text{-}DSPACE(L_D(n))$
(3) $weak\text{-}NSPACE(L_N(n)) \neq weak\text{-}NSPACE(L_N(n))$
(4) $L_D(n)$ and $L_N(n)$ are not fully space constructible.

Proof. In Section 7.5 it was shown that there exist computable functions $L_D(n)$ and $L_N(n)$, such that $L_1(n) \leq L_D(n)$, $L_N(n) \leq L_2(n)$, and the classes

$$weak\text{-}DSPACE(L_D(n)) \quad \text{and} \quad weak\text{-}NSPACE(L_N(n))$$

are not closed under complement.

Since, by Theorem 7.2.2, the class $strong\text{-}DSPACE(L_D(n))$ is closed under complement,

$$weak\text{-}DSPACE(L_D(n)) \neq strong\text{-}DSPACE(L_D(n)).$$

To prove (3), let us consider the language $A \subset \{1\}^*$ described in the proof of Theorem 7.5.4. It was proved that

$$A \in weak\text{-}NSPACE(L_N(n))$$

and its complement

$$A^c \notin weak\text{-}NSPACE(L_N(n)).$$

Hence, also

$$A^c \notin strong\text{-}NSPACE(L_N(n))$$

and, by Theorem 7.4.2,

$$A \notin strong\text{-}NSPACE(L_N(n)).$$

(4) follows immediately from (2), (3), and Theorem 8.1.1. $\square$

8.3. Weak and Strong Complexity Classes below Logarithm

Now we shall show that if $L(n) \geq \log\log n$ and $L(n) = o(\log n)$, then deterministic, nondeterministic, and alternating weak $L(n)$ space complexity classes are not equal to the strong ones.

Theorem 8.3.1. If $L(n) \geq \log\log n$ and $L(n) = o(\log n)$, then

$$strong\text{-}DSPACE[L(n)] \neq weak\text{-}DSPACE[L(n)],$$
$$strong\text{-}NSPACE[L(n)] \neq weak\text{-}NSPACE[L(n)], \quad and$$
$$strong\text{-}ASPACE[L(n)] \neq weak\text{-}ASPACE[L(n)],$$

Proof. Consider the language

$$A = \{a^k b^m \mid k \neq m\}.$$

By Lemma 4.1.3,

$$A \in weak\text{-}DSPACE[\log\log n].$$

On the other hand, in Lemma 8.3.3 below we shall show that if $L(n) = o(\log n)$, then A is not in $strong\text{-}ASPACE[L(n)]$.   □

Note that it is easy to prove that $A \notin strong\text{-}NSPACE[o(\log n)]$, one should only use Theorem 7.4.2 and Corollary 6.2.5. To prove that A is not in $strong\text{-}ASPACE[L(n)]$ we need the following:

Lemma 8.3.2. Let M be an alternating Turing machine accepting the language $\{a^k b^m \mid k \neq m\}$ or $\{a^k b^k \mid k \geq 0\}$. Then on almost every (except possibly finitely many) input of the form $a^k b^k$ there is a computation of M which uses at least $c \cdot \log\log\log n$ space for some constant $c > 0$.

Proof. We assume that whenever M stops it stops on the right endmarker. Let C_k denotes the set of internal configurations of M on the input $a^k b^k$. For each j, $j \leq k$, consider the behavior of M on the input $a^k b^j b^{k-j} = a^k b^k$.

Let β be an internal configuration and i be a position on the input w. Then a *subtree* T rooted in β at i is a finite tree whose nodes are labeled by configurations of M on w. The root of the tree is labeled by β positioned at cell i, the children of any node labeled by a universal configuration are all immediate successors of that configuration; and any node labeled by an existential configuration has one child, which is labeled by one of the immediate successors of that configuration (provided there are any).

Let T be a subtree rooted in an internal configuration β at the first or at the jth b. Consider the set of all internal configurations of T encountered just after M leaves the first b^j sector for the first time. And define $SEC(T)$ in the following way: for any internal configuration γ and $s \in \{l, r\}$, a pair $(s, \gamma) \in SEC(T)$ if there is a path in the tree T which leaves b^j for the first time in the internal configuration γ at the left end, if $s = l$, or at the right end, if $s = r$.

Now we define the following relation:

$$R(j) \subseteq \{l, r\} \times C_k \times \mathcal{P}(\{l, r\} \times C_k)$$

$(s, \beta, B) \in R(j)$ iff there is a subtree T rooted at β at the first b, if $s = l$, or at the jth b, if $s = r$, such that $B = SEC(T)$. $\mathcal{P}(A)$ denotes the set of all subsets of A.

It is obvious that for any two numbers $i < j \leq k$, $R(i) \neq R(j)$, because, otherwise, M would not be able to distinguish $a^k b^k = a^k b^i b^{k-i}$ from $a^k b^j b^{k-i}$. So, the number of possible relations $|\{R(j) \mid j \leq k\}|$ should be not less than k. But

$$|\{R(j) \mid j \leq k\}| \leq 2^{2|C_k|2^{2|C_k|}} \leq 2^{d|C_k|}$$

for some constant $d > 0$, and, by Lemma 2.3.2, the number of possible internal configurations $|C_k| \leq d_1^{s_k}$, where s_k is the maximal space used by M on $a^k b^k$ and d_1 is a constant. So, we have

$$2^{d|C_k|} \geq k \quad \text{and} \quad |C_k| \leq d_1^{s_k}.$$

And there exists a constant $c > 0$, such that

$$s_k \geq c \log \log \log k,$$

for almost every k (except possibly finitely many). Hence, for almost every k, the maximal space used by M on $a^k b^k$ is not less than $c \cdot \log \log \log k$.   □

Now we can prove that neither $\{a^k b^m \mid k \neq m\}$ nor $\{a^k b^k \mid k \geq 0\}$ can be accepted by a strongly $o(\log n)$ space bounded alternating Turing machine.

Lemma 8.3.3. If $L(n) = o(\log n)$, then

$$\{a^k b^m \mid k \neq m\} \notin strong\text{-}ASPACE[L(n)] \quad \text{and}$$
$$\{a^k b^k \mid k \geq 0\} \notin strong\text{-}ASPACE[L(n)]$$

Proof. Suppose, for a contradiction, that there exists a strongly $L(n)$ space bounded alternating Turing machine M which accepts $\{a^k b^m \mid k \neq m\}$ or $\{a^k b^k \mid k \geq 0\}$. Then, by Lemma 8.3.2, on almost every input of the form $a^k b^k$ there is a computation of M which uses at least $c \log \log \log k$ space for some constant $c > 0$. But then there is also a strongly $L(n)$ space bounded nondeterministic Turing machine M_1 which uses at least $c \log \log \log k$ space on almost every input of the form $a^k b^k$. M_1 works exactly like the alternating Turing machine M, but all its states are existential.

Now we shall show that then there exists a strongly $L_1(n) = L(2n)$ space-bounded nondeterministic Turing machine M_2 which uses at least $\log \log \log n$ space on every input of length n. When acting on any input of length n, M_2

simulates M_1 working on $a^n b^n$. During the simulation M_2 keeps in its memory in which sector, a-s or b-s, the simulated M_1 acts and accordingly reads all input symbols as a or b. First, reading all symbols as a, M_2 simulates M_1 working on the a^n sector. When M_1 reaches the end of a^n then M_2 reaches the end of the input, and when M_1 enters the b^n sector, then M_2 switches itself to read b and simulates M_1 in b^n changing only the directions of the moves, i.e. when M_1 moves left then M_2 moves right and vice versa. Note that $L_1(n) = o(\log n)$.

But, on the other hand, in the proof of Theorem 6.3.1 it was shown that if $L(n) = o(\log n)$ or even if $\liminf \frac{L(n)}{\log n} = 0$, then for every strongly $L(n)$ space-bounded nondeterministic Turing machine M, there exists a constant c, such that M uses at most c space on infinitely many inputs. □

Note that the language A can be accepted by a strongly $L(n)$ space-bounded deterministic Turing machine with $L(n) \ll \log n$ (i.e. $\liminf \frac{L(n)}{\log n} = 0$). Consider the function

$$L(n) = \begin{cases} 0, & \text{for } n \text{ odd,} \\ \log n, & \text{for } n \text{ even.} \end{cases}$$

Obviously $L(n) \ll \log n$ and there is a strongly $L(n)$ space-bounded deterministic Turing machine M accepting A. M first checks if the length of an input is odd, accepts if it is, and if not, counts a's and b's, and compares the numbers.

We shall show that Theorem 8.3.1 is also valid if $\log \log n \leq L(n) \ll \log n$.

Theorem 8.3.4. If $L(n) \geq \log \log n$ and $\liminf \frac{L(n)}{\log n} = 0$, then

$$strong\text{-}DSPACE[L(n)] \neq weak\text{-}DSPACE[L(n)],$$
$$strong\text{-}NSPACE[L(n)] \neq weak\text{-}NSPACE[L(n)], \text{ and}$$
$$strong\text{-}ASPACE[L(n)] \neq weak\text{-}ASPACE[L(n)],$$

Proof. Consider the language

$$C = \{a^k b^m \mid m \neq k \text{ and } m \neq k - 1\}.$$

Similarly to the proof of Lemma 4.1.3 one can prove that

$$C \in weak\text{-}DSPACE[\log \log n],$$

and similarly to the proof of Lemma 8.3.3 one can show that

$$C \notin strong\text{-}ASPACE[L(n)].$$

□

8.4. Strong Mode of Space Complexity for Recognizing Machines

A recognizing Turing machine has two special states s_a and s_r and reaches s_a on every accepted input, and s_r on every not accepted one, and on no input both s_a and s_r can be reached. Observe that, to prove that a deterministic recognizing Turing machine M is strongly $L(n)$ space-bounded, it suffice to show that there is at least one $L(|w|)$ space-bounded computation for every input w. Now we show that the same is true for nondeterministic recognizing Turing machines with logarithmic or greater space. Namely,

Theorem 8.4.1.   For arbitrary function $L(n) \geq \log n$, if a language A is accepted by a recognizing Turing machine M which has at least one accepting (ending with s_a) or rejecting (ending with s_r) computation on every input w, then A is also accepted by strongly $L(n)$ space-bounded recognizing Turing machine.

Proof.   We shall show that M can be simulated by a strongly $L(n)$ space-bounded recognizing nondeterministic Turing machine M_R. On any input w, M_R marks off $s = \log(|w|)$ cells on the work tape and using Immerman-Szelepcsényi techniques (see Theorem 7.3.1) counts all s space-bounded accessible configurations of M on w. If among these configuration is an accepting one (with s_a), then M_R accepts, if there is a rejecting (with s_r), then it rejects, and if M_R does not find neither any accepting nor any rejecting one, then it marks one cell more $s := s + 1$ and repeats the process. □

8.5. Weak and Middle Modes of Space Complexity above Logarithm

Observe that the middle and weak modes of space complexity are equivalent for deterministic Turing machines. Now we shall show that the same is true for nondeterministic Turing machines with logarithmic or greater space.

Theorem 8.5.1.   If $L(n) \geq \log n$ then weak and middle modes of space complexity for $L(n)$ space-bounded nondeterministic recognizing Turing machines are equivalent.

Proof.   Similar to the proof of Theorem 8.4.1. The Turing machine M_R counts s space-bounded accessible configurations and looks for an accepting one. □

9. Padding

The question whether the nondeterministic and deterministic $L(n)$ space complexity classes are equal is a major open question in the space complexity theory. The answer to this question is known only for some special functions $L(n)$ and models of computation. As we have already mentioned, if $L(n) = o(\log \log n)$, then deterministic, nondeterministic, and even alternating Turing machines accept only regular languages. One can also construct, using diagonalization, a total recursive function $L(n)$ such that the nondeterministic and deterministic $L(n)$ space complexity classes are equal (Hopcroft and Ullman 1979).

In this chapter we show that the question whether the nondeterministic and deterministic space complexity classes are equal for "large" function can be reduced to the analogous question for "small" function. Using a padding argument of (Savitch 1970) we shall prove, that for any "well-behaved" function $L(n) \geq \log n$, if the deterministic and nondeterministic space complexity classes are equal for space bound $L(n)$, then they are also equal for any space-bound function $H(n) \geq L(n)$. We shall also present the result of (Szepietowski 1990) which says that if the deterministic and nondeterministic space complexity classes are equal for space bound $\log \log n$, then they are also equal for $\log n$. Moreover, we show that there exists a language $H \in NSPACE[\log \log n]$ such that $H \in DSPACE[\log \log n]$ if and only if $DSPACE[\log n] = NSPACE[\log n]$.

9.1. Padding above Logarithm

In this section we show that if the deterministic and nondeterministic space complexity classes are equal for some fully space constructible $L(n) \geq \log n$, then they are also equal for any space-bound function $H(n) \geq L(n)$ (Savitch 1970).

Theorem 9.1.1. Let $L(n)$ be fully space constructible function $L(n) \geq \log n$ and $c_1 \leq \frac{L(n)}{L(n+1)} \leq c_2$ for all n and some constants $c_1, c_2 > 0$. Then for every function $H(n) \geq L(n)$

$$DSPACE[L(n)] = NSPACE[L(n)]$$
$$\Downarrow$$
$$DSPACE[H(n)] = NSPACE[H(n)].$$

The theorem is valid for both the strong and weak mode of space complexity.

Note that the above assumptions are valid if $L(n)$ is any polynomial in $\log n$, n, or d^n for some constant d, and we have, for example, for every $H(n) \geq \log n$.

$$DSPACE[\log(n)] = NSPACE[\log(n)]$$
$$\Downarrow$$
$$DSPACE[H(n)] = NSPACE[H(n)].$$

Proof of Theorem 9.1.1. First we show that the theorem is valid for weak mode of space complexity.

Suppose that a language $G \subseteq \Sigma^*$ is accepted by a weakly $H(n)$ space-bounded Turing machine M. Let us define so-called *padding* of G:

$$Pad(G, M) = \{xb^m \mid x \in G \text{ and } x \text{ is accepted by } M \text{ within } L(|x| + m) \text{ space}\},$$

where b is an additional symbol, $b \notin \Sigma$.

First we show that

$$Pad(G, M) \in NSPACE[L(n)]$$

The $L(n)$ space-bounded nondeterministic Turing machine M_N accepting $Pad(G, M)$ works as follows. On any input w of the form $w = xb^m$, M^P first marks off $L(|w|) = L(|x| + m)$ space (this is possible, since $L(n)$ is fully space constructible). M_N will not use more than $L(|x| + m)$ space. Then M_N checks if M accepts x within marked space (see Theorem 7.3.3) and accepts if it does.

And now by our assumption,

$$Pad(G, M) \in DSPACE[L(n)]$$

and there exists an $L(n)$ space-bounded deterministic Turing machine M_D accepting $Pad(G, M)$. We shall show that there also exists a weakly $H(n)$ space-bounded deterministic Turing machine M_1 accepting G. On an input x, M_1 first, assumes $m = 1$ and simulates M_D on xb^m if M_D accepts xb^m within $L(|x| + m)$, then M_1 accepts x, if not, M_1 assumes $m = 2$, and so on. The number m is remembered in binary and M_1 while simulating M_D on xb^m keeps in binary the position of the input head when it is within b^m. This can be done within $L(|x| + m)$ space, since $\log m \leq \log(|x| + m) \leq L(|x| + m)$.

Now we shall show that M_1 accepts G and is weakly $dH(n)$ space-bounded for some constant d. Suppose that $x \in G$, then $xb^m \in Pad(G, M)$ for some sufficiently large m, and let m be the first such number. Then M_1 accepts x using $L(|x| + m)$ space.

Observe that

$$L(|x| + m - 1) \leq H(|x|).$$

(and also $L(|x| + k) \leq H(|x|)$ for every $k < m$) for otherwise if $L(|x| + m - 1) > H(|x|)$, then x is accepted by M within $L(|x| + m - 1)$ space and $xb^{m-1} \in Pad(G, M)$, which contradicts the minimality of m. On the other hand, by the assumption,

$$c_1 L(|x| + m) \leq L(|x| + m - 1)$$

and there is a constant d such that

$$L(|x| + m) \leq dH(|x|).$$

Thus, x is accepted by M_1 within $dH(n)$ space. If $x \notin G$ then for any m, $xb^m \notin Pad(G, M)$ and x is not accepted by M_1.

Hence, we have shown that if $DSPACE[L(n)] = NSPACE[L(n)]$ then the language G is also accepted by a deterministic weakly $H(n)$ space-bounded Turing machine. This completes the proof for weak mode of space complexity.

To prove the theorem for strong mode of space complexity we shall need more sophisticated methods. The main idea of the proof is the same as before, but some details should be added.

Now G is accepted by a strongly $H(n)$ space-bounded Turing machine M and by Theorem 7.3.1 we can assume that M is recognizing, i.e. it has two states s_a and s_r and reaches the state s_a on every $x \in G$, and state s_r on every $x \notin G$ (and on no input both s_a and s_r can be reached). The padding of G is defined as follows.

$Pad(G, M) = \{xb^m \mid x \in G$ and x is accepted by M within $L(|x| + m)$ or
$L(|x| + m - 1)$ space, or
$x \notin G$ and M reaches the state s_r within $L(|x| + m)$ and
not within $L(|x| + m - 1)$ space$\}$

The strongly $L(n)$ space-bounded nondeterministic Turing machine M_N which accepts $Pad(G, M)$ works as follows. First using the Immerman-Szelepcsényi technique (see Theorem 7.3.3) M_N checks whether the state s_a or s_r can be reached by M on x within $L(|x| + m)$ or $L(|x| + m - 1)$ space, and accepts if s_a can be reached within $L(|x| + m)$ or $L(|x| + m - 1)$ space, or if s_r can be reached within $L(|x| + m)$ but not within $L(|x| + m - 1)$ space. M_1 can be made strongly $L(n)$ space bounded, because, by the assumption $L(|x| + m - 1) \leq c_2 L(|x| + m)$.

By our assumption, $Pad(G, M)$ is also accepted by a $L(n)$ space-bounded deterministic Turing machine M_D. And we shall show that G can be also accepted by a strongly $H(n)$ space-bounded Turing machine M_1 which works as follows. On an input x, M_1 looks for the first m such that $xb^m \in Pad(G, M)$. Note that for every input x such m exists. Then M_1 checks using $L(|x| + m + 1)$ space if $xb^{m+1} \in Pad(G, M)$, and accepts x if $xb^{m+1} \in Pad(G, M)$, and reject, if not.

In much the same way as above can show that there is a constant d such that M_1 reaches the state s_a or s_r within $dH(n)$ space on every input.   □

9.2. Padding below Logarithm

In this section we present the result of (Szepietowski 1990) that

$$DSPACE[\log \log n] = NSPACE[\log \log n]$$
$$\Downarrow$$
$$DSPACE[\log n] = NSPACE[\log n].$$

This is valid for both the weak and strong mode of space complexity.

For an arbitrary language G over an alphabet Σ, we define a new *padding* of G:

$$P(G) = \{xb^m \mid x \in G \text{ and } m \text{ is a common multiple of all } i \leq |x|\},$$

where b is an additional symbol, $b \notin \Sigma$.

First, we shall show that if G is accepted within $\log n$ space, then $P(G)$ is accepted within $\log \log n$ space.

Lemma 9.2.1.

If $G \in XSPACE[\log n]$ then $P(G) \in strong\text{-}XSPACE[\log \log n]$,

where X stands for D or N.

Proof. Let M be $\log n$ space-bounded deterministic or nondeterministic Turing machine accepting G. By Theorem 8.1.1, we can assume that M is strongly space-bounded because $\log n$ is fully space constructible. Then there is strongly $\log \log n$ space-bounded deterministic or nondeterministic, respectively, Turing machine M^P accepting $P(G)$. On any input xb^m, M^P checks if m is a common multiple of all $i \leq |x|$. To do this, it computes the binary description of

$$F(m) = \min\{i \mid i \text{ does not divide } m\}$$

and compares it with $|x|$. If $F(m) \leq |x|$, then m is not divisible by some $i \leq |x|$ and the input is rejected. Otherwise, M^P simulates M on x. By Lemma 4.1.2,

$$F(m) \leq c \log m \qquad \text{for some constant } c,$$

and if m is a common multiple of all $i \leq |x|$, then by Lemma 4.1.2, there is a constant $c_1 > 1$ such that
$$m \geq c_1^{|x|}.$$

Hence, the space used by M is bounded by

$$\log |x| \leq d \log \log m \leq d \log \log |w|$$

for some constant d. Then M^P is strongly $\log \log n$ space-bounded.   □

In the following lemma we show that if $P(G) \in DSPACE[\log \log n]$, then $G \in DSPACE[\log n]$. This will suffice for our purpose. However, using the result of (Geffert 1991), one can also show that the same is true for nondeterministic space.

Lemma 9.2.2.

If $P(G) \in weak\text{-}DSPACE[\log \log n]$ then $G \in DSPACE[\log n]$.

Proof. Let M^P be a weakly $\log \log n$ space-bounded deterministic Turing machine accepting $P(G)$. We shall show that there exists a weakly $\log n$ space-bounded deterministic Turing machine M accepting G. Since $\log n$ is fully space constructible, this means that G can also be accepted by a strongly $\log n$ space-bounded deterministic Turing machine. If a word $x \in \Sigma^*$ is presented to M, it simulates M^P working on the word xb^m, where m is the least common multiple of all $i \le |x|$. We denote this word by $P(x)$. Note that

$$P(x) \in P(G) \quad \text{if and only if} \quad x \in G.$$

We shall show that M accepts x if and only if M^P accepts $P(x)$, and that M uses on x the same amount of space as M^P on $P(x)$.

While (the input head of) M^P is on x, M simply acts like M^P. When M^P enters the b^m sector, four cases are possible (we assume that M^P enters b^m from x, the other case is simulated analogously):

(a) M^P loops by repeating its internal configuration on b^m,
(b) M^P stops inside b^m,
(c) M^P leaves b^m before it has repeated its internal configuration,
(d) M^P neither loops nor stops, nor leaves b^m.

Let β_t and d_t denote the internal configuration and distance from x after t steps of M^P in b^m. Since M^P is deterministic, the internal configuration β_t and the move $d_t - d_{t-1}$ are determinated by β_{t-1}, provided that after $t-1$ steps M^P is still inside b^m.

At the beginning M simulates the first steps of M^P inside b^m up to the moment when either M^P repeats its internal configurations, stops, or leaves b^m. During these first steps M keeps in its memory the current values of t, d_t, and β_t, and after simulating the t-th step of M^P, M checks if β_t is equal to some β_j with $j < t$ (it remembers t, d_t, and β_t, and reruns the computation in b^m). In the case (d), both M^P and M never stop using more and more space and neither of them accepts its input. In the other cases, space used by M^P inside b^m is bounded by, say k. By Lemma 2.3.2, the number of all internal configurations of M^P inside b^m is then bounded by c^k for some constant $c > 1$, and M^P can perform at most c^k steps without repeating its internal configuration. So, using k space M can remember the current values of t, d_t, and β_t up to the moment when M^P either repeats its internal configuration, stops, or leaves b^m.

In the case (b) or (c), M simulates all steps of M^P inside b^m. In the case (a), M computes the internal configuration of M^P after leaving b^m or detects that M^P does not leave b^m. If M^P repeats its internal configuration inside b^m, then there are two positive integers q, r such that $\beta_q = \beta_{q+r}$. We can assume that $q + r < c^k$. If $d_q = d_{q+r}$, then M^P loops and never leaves b^m. This can be detected, since M keeps in the memory current d_q. If $d_q > d_{q+r}$, then M^P comes back to x after no more than c^{2k} steps. All these steps can be simulated by M in k space and M can then compute the internal configuration of M^P after leaving b^m.

If $d_q < d_{q+r}$, then the internal configuration of M^P loops and after each loop M^P is farther from x, and it reaches the second end of b^m after some steps. In this case, M computes $m(\text{mod } d)$, where $d = d_{q+r} - d_q$ is the distance made by M^P during one loop. By Lemma 4.1.2, if M is the least common multiple of all numbers $i \leq |x|$, then

$$m = \prod_{\substack{all\ primes\ p}} p^{\lfloor \log_p |x| \rfloor}.$$

Hence,

$$m(\text{mod } d) = \Big(\prod_{\substack{all\ primes\ p}} p^{\lfloor \log_p |x| \rfloor}(\text{mod } d)\Big) (\text{mod } d).$$

M_p puts 1 on a counter modulo d and then one by one:

- finds p_i –the i-th prime not greater than $|x|$,
- computes p_i^y such that $p_i^y \leq |x| < p_i^{y+1}$, note that $y = \lfloor \log_p |x| \rfloor$,
- computes $p_i^y(\text{mod } d)$, and
- multiplies the contents of the counter by $p_i^y(\text{mod } d)$.

This can be done in k space because $d \leq c^k$. Observe that $p^{\lfloor \log_p |x| \rfloor} = 1$ for every $p > |x|$. Next M computes

$$z = (m(\text{mod } d) - d_q - r)(\text{mod } d),$$

where r is the number of steps during one loop, and stands at the u-th b with

$$u = m - r - z,$$

provided that $m - r - z \geq d_q$. Note that

$$u = d_q(\text{mod } d),$$

so there is a moment, say v, that M^P visits the u-th b with the configuration β_q. Since $u \leq m - r$, M^P does not leave b^m before v. Now M simulates steps of M^P from the moment v, i.e. from the u-th b and the internal configuration $\beta_v = \beta_q$, up to the moment it leaves b^m for the first time. During the simulation M keeps in its memory the distance between current position of M^P and the far end of b^m. At the moment v this distance is equal to $r + z$ and since $r + z \leq c^k$, it can be remembered using k space. If $m - r - z < d_q$, then all steps of M^P inside b^m are simulated.

It is easy to see that M accepts x if and only if M^P accepts $P(x)$ and that M uses on x the same amount of space as M^P on $P(x)$. We shall now show that M is weakly $\log n$ space-bounded. If $x \in G$, then space used by M on x is equal to the space used by M^P on $P(x)$ and does not exceed $\log \log |xb^m|$. By Lemma 4.1.2, $m \leq c_2^{|x|}$, so

$$\log \log |xb^m| \leq \log \log(2m) \leq c \log |x|$$

for some constant c. Hence, M is weakly $\log n$ space-bounded and since $\log n$ is fully space constructible, G can also be accepted by a strongly $\log n$ space-bounded deterministic Turing machine.   □

Now we shall prove the following lemma.

Lemma 9.2.3.
$$strong\text{-}NSPACE[\log \log n] \subseteq weak\text{-}DSPACE[\log \log n]$$
$$\Downarrow$$
$$DSPACE[\log n] = NSPACE[\log n].$$

Let us remember that $\log n$ is fully space constructible and it does not matter which mode of space complexity is used for $\log n$.

Proof. Suppose that
$$strong\text{-}NSPACE[\log \log n] \subseteq weak\text{-}DSPACE[\log \log n]$$
and that G is an arbitrary language such that
$$G \in NSPACE[\log n],$$
then by Lemma 9.2.1,
$$P(G) \in strong\text{-}NSPACE[\log \log n],$$
and by the assumption,
$$P(G) \in weak\text{-}DSPACE[\log \log n]$$
and by Lemma 9.2.2,
$$G \in DSPACE[\log n].$$
Thus, we have shown that
$$strong\text{-}NSPACE[\log \log n] \subseteq weak\text{-}DSPACE[\log \log n]$$
$$\Downarrow$$
$$NSPACE[\log n] \subseteq DSPACE[\log n].$$
Lemma 9.2.3 follows from this and the trivial fact that
$$NSPACE[\log n] \supseteq DSPACE[\log n].$$

□

Now we are ready to present the main theorem of this section.

Theorem 9.2.4.

$$DSPACE[\log \log n] = NSPACE[\log \log n]$$
$$\Downarrow$$
$$DSPACE[\log n] = NSPACE[\log n].$$

The theorem is valid for both the strong and weak mode of space complexity.

Proof.   It is trivial that

$$strong\text{-}NSPACE[\log \log n] \subseteq weak\text{-}NSPACE[\log \log n]$$

and

$$strong\text{-}DSPACE[\log \log n] \subseteq weak\text{-}DSPACE[\log \log n].$$

Thus, if

$$strong\text{-}NSPACE[\log \log n] = strong\text{-}DSPACE[\log \log n]$$

or

$$weak\text{-}NSPACE[\log \log n] = weak\text{-}DSPACE[\log \log n],$$

then

$$strong\text{-}NSPACE[\log \log n] \subseteq weak\text{-}DSPACE[\log \log n]$$

and by Lemma 9.2.3,

$$NSPACE[\log n] = DSPACE[\log n].$$

$$\square$$

Also any middle mode of space complexity can be used, as long as

$$strong\text{-}NSPACE[\log \log n] \subseteq middle\text{-}NSPACE[\log \log n] \quad \text{and}$$
$$middle\text{-}DSPACE[\log \log n] \subseteq weak\text{-}DSPACE[\log \log n].$$

For example, the middle mode defined in Definition 2.5.3, where it is assumed that all computations on all accepted inputs satisfy the space bound. Thus, we have.

Corollary 9.2.5.   If any middle mode of space complexity satisfies the above conditions, then

$$middle\text{-}NSPACE[\log \log n] = middle\text{-}DSPACE[\log \log n]$$
$$\Downarrow$$
$$DSPACE[\log n] = NSPACE[\log n].$$

At the end we prove the following theorem.

Theorem 9.2.6. There exists a language $H \in strong\text{-}NSPACE[\log \log n]$ such that $H \in weak\text{-}DSPACE[\log \log n]$ if and only if

$$DSPACE[\log n] = NSPACE[\log n].$$

Proof. Consider an arbitrary language G that is complete for $NSPACE[\log n]$ (see Section 3.3). Thus, we have

(1) $G \in NSPACE[\log n]$ and
(2) $G \in DSPACE[\log n]$ if and only if $DSPACE[\log n] = NSPACE[\log n]$.

Let

$$H = P(G),$$

then by (1) and Lemma 9.2.1,

$$H \in strong\text{-}NSPACE[\log \log n].$$

Suppose now that

$$H \in weak\text{-}DSPACE[\log \log n].$$

Then by Lemma 9.2.2,

$$G \in DSPACE[\log n]$$

and by (2),

$$DSPACE[\log n] = NSPACE[\log n].$$

On the other hand, if

$$DSPACE[\log n] = NSPACE[\log n],$$

then by (2),

$$G \in DSPACE[\log n]$$

and by Lemma 9.2.1,

$$H \in strong\text{-}DSPACE[\log \log n].$$

$\square$

10. Deterministic Versus Nondeterministic Turing Machines

In this chapter we consider the cost of determinization of nondeterministic Turing machines with bounded space. More precisely, we shall try to find out how much space is needed for the deterministic Turing machine to accept the language which is accepted by a nondeterministic Turing machine in $L(n)$ space. We present the famous result of Savitch (1970), who proved that if $L(n) \geq \log n$, then every $L(n)$ space-bounded nondeterministic Turing machine can be simulated by an $[L(n)]^2$ space-bounded deterministic Turing machine. Monien and Sudborough (1982) proved that if $L(n) \leq \log n$ and $L(n)$ is fully space constructible or $L(n) = f(\log n)$ for some fully space constructible $f(n)$, then every $L(n)$ space-bounded nondeterministic Turing machine can be simulated by an $L(n) \cdot \log n$ space-bounded deterministic Turing machine. In particular, every $\log \log n$ space-bounded nondeterministic Turing machine can be simulated by a $\log n \cdot \log \log n$ space-bounded deterministic Turing machine. From the result of Geffert (1991) it follows that for an arbitrary $L(n)$, if $A \subseteq a_1^* a_2^* \ldots a_k^*$ and A is accepted by a strongly $L(n)$ space-bounded nondeterministic Turing machine, then it can also be accepted by a strongly $[L(n)]^2$ space-bounded deterministic Turing machine.

10.1. Determinism Versus Nondeterminism above Logarithm

In this section we present the result of Savitch (1970).

Theorem 10.1.1. If $L(n) \geq \log n$, then

$$NSPACE[L(n)] \subseteq DSPACE[[L(n)]^2].$$

The theorem is valid for both the strong and weak mode of space complexity.

Proof. Suppose first that $L(n)$ is fully space constructible. Let M be an $L(n)$ space-bounded nondeterministic Turing machine and w be an input of length n. We present an $[L(n)]^2$ space-bounded deterministic Turing machine M^D that accepts the same language as M. Consider the set of all $L(n)$ space-bounded configurations of M on w. By Lemma 2.2.2, there are no more than $s(n + 2)L(n)t^{L(n)}$ configurations, and since $L(n) \geq \log n$, there exists a constant d such that $s(n+2)L(n)t^{L(n)} \leq d^{L(n)}$ and d depends only on s and t. Thus, M^D can keep in its memory numbers up to the number of all configurations of M. To recognize if M accepts w, M^D has to find out if there is a computation of M not longer than $d^{L(n)}$ from the initial configuration to an accepting configuration. This is done by a divide-and-conquer algorithm. To find out if there is a computation from the initial configuration to an accepting configuration, M^D performs recursively the following procedure: To find out if there is a computation not longer than m from a configuration β_1 to a configuration β_2, M^D checks for each configuration β_3 if

- there is a computation not longer than $\frac{m}{2}$ from β_1 to β_3 and
- there is a computation not longer than $\frac{m}{2}$ from β_3 to β_2.

This algorithm keeps in memory the list of $\log m$ configurations of M, where m is the length of the accepting computation of M on w. Since $m \leq d^{L(n)}$, it needs $[L(n)]^2$ space.

If $L(n)$ is not fully space constructible, M^D first assumes $L(n) = \log n$. If M accepts within space $L(n) = \log n$, then M^D accepts. If M does not accept within space $\log n$, then M^D assumes $L(n) = \log n + 1$, and so on. If M accepts the input w, then M^D will eventually find the correct value of $L(n)$ and accept within $[L(n)]^2$ space. Hence,

$$weak\text{-}NSPACE[L(n)] \subseteq weak\text{-}DSPACE[[L(n)]^2].$$

If we use the strong mode of space complexity, then M^D before assuming $L(n) = k + 1$ first checks if M uses $k + 1$ space on w. To do this, M^D checks for each k space-bounded configuration β which is about to use $k + 1$ space in the next step, if there is a k space-bounded computation from the initial configuration to β. Thus, the theorem is also valid for the strong mode of space complexity.

$$strong\text{-}NSPACE[L(n)] \subseteq strong\text{-}DSPACE[[L(n)]^2].$$

$\square$

10.2. Determinism Versus Nondeterminism below Logarithm

First, we present a result of Monien and Sudborough (1982).

Theorem 10.2.1. If $L(n) \leq \log n$ and

- $L(n)$ is fully space constructible or
- $L(n) = f(\log n)$ for some fully space constructible $f(n)$,

then

$$NSPACE[L(n)] \subseteq DSPACE[L(n) \cdot \log n].$$

In particular,

$$NSPACE[\log \log n] \subseteq DSPACE[\log n \cdot \log \log n].$$

Proof. Let M be an $L(n)$ space-bounded nondeterministic Turing machine and w be an input of length n. We present an $L(n) \cdot \log n$ space-bounded deterministic Turing machine M^D that accepts the same language as M. Let

$$\beta_1, \beta_2, \ldots, \beta_m \qquad (1)$$

be a sequence of all $L(n)$ space-bounded configurations of M ordered in such a way that configurations with input head positions i stand before configurations with input head positions $i+1$. Thus, for any two configurations β_j and β_k, if β_k can be reached from β_j in one step, then the distance between β_j and β_k in the sequence (1) is not longer than the number of all $L(n)$ space-bounded internal configurations of M, i.e. $|k - j| \le d^{L(n)}$, for some constant d. This fact was used to design a more economical divide-and-conquer algorithm.

Suppose that β_j and β_k are two configuration in the sequence (1) and that M^D has to check if there is a computation of M from β_j to β_k. Let W be a so-called window of size $2^{L(n)}$ in the middle of the sequence

$$\beta_j, \beta_{j+1}, \ldots, \beta_k.$$

The window W consists of configurations from $\frac{i+k}{2} - 2^{L(n)}$ to $\frac{i+k}{2} + 2^{L(n)}$. If there is a computation from β_j to β_k, then it must contain some configurations from W.

To find out if there is a computation of M from the initial configuration to an accepting one, M^D performs recursively the following: To find out if there is a computation from β_j to β_k for $|k - j| > 2^{L(n)}$, M^D checks for all configuration β from the window W if

- there is a computation from β_j to β and
- there is a computation from β to β_k.

If $|k - j| \le 2^{L(n)}$, M^D proceeds like in Savitch's algorithm. M^D remembers only the distance between β and the center of the window W which is not longer than $2^{L(n)}$ and requires $L(n)$ space. M^D needs to remember a list of such distances. The exact position of a configuration can be computed from its distance from the center of the window. The position of the center of a window can be computed from the position of the configuration on the list. The length of the list is bounded by $O(\log n)$ because the number of all configurations of M on w is bounded by $cnd^{L(n)} \le n^r$ for some constants c and r, and after each stage of recursion the distance between β_j and β_k is halved.

□

Geffert (1991) noticed (without any proof) that to bounded languages (i.e. to the languages of the form $A \subseteq a_1^* a_2^* \ldots a_k^*$) a modified divide-and-conquer algorithm can be applied.

Theorem 10.2.2. Let $A \subseteq a_1^* a_2^* \ldots a_k^*$ and $L(n)$ be an arbitrary function, then

if $A \in$ *strong-NSPACE*$[L(n)]$ then $A \in$ *strong-DSPACE*$[(L(n))^2]$.

Proof. We shall only prove the result for languages of the form $A \subseteq a^* b^*$, but the proof can be easily generalized for the other cases.

The proof goes similarly as in the proof of Theorem 7.4.1, we shall only consider configurations with positions of the input head in the inshore sectors, i.e. at a distance of at most $[r(n)]^4$ from either of endmarkers or from the boundary between a's and b's, where $r(n)$ denotes the number of all possible $L(n)$ space-bounded internal configurations of M. By Lemma 2.3.2, there is a constant d_1 such that $r(n) \leq d_1^{L(n)}$.

We apply Savitch's divide-and-conquer algorithm to all such configurations. Note that the number of all configurations with the input head in the inshore sectors is not greater than $d^{L(n)}$ for some constant d, and can be remembered using $L(n)$ space. Similarly, any configuration from the inshore sectors can be kept in the memory. In this case only the distance between the input head and the nearest endmarker or boundary is remembered.

We modify the notion of immediate successor of a configuration staying in the inshore sectors. We say that a configuration β' is an immediate successor of another configuration β also if β has its input head position at the distance of $[r(n)]^4$ from one end of a's (or b's); β' has its input head position at the distance of $[r(n)]^4$ from the other end of a's (or b's, respectively) and β' can be reached from β without going outside a's (or b's).

From the proof of Lemma 6.3.3 it follows that it suffices to consider only configuration that can be reached by looping (i.e. by repeating internal configuration) because if there is an accepting computation of M on w, then there is also an accepting computation that always loops through the deep see sectors of a's and b's, i.e. through a's and b's that are further than $[r(n)]^4$ from the endmarkers or the boundary between a's and b's.

In order to check if, for example, β' positioned at $k - [r(n)]^4$ in a^k can be reached by looping from β positioned at $[r(n)]^4$, M^D checks one by one for each pair of numbers i and j, $1 \leq i < j \leq r(n)$, if there are two configurations β_1 and β_2 such that:

- the input head position of β_1 is equal to $[r(n)]^4 + i$,
- the input head position of β_2 is equal to $[r(n)]^4 + j$,
- both the internal configurations of β_1 and β_2 are equal to the internal configuration of β',
- β_1 can be reached from β in less than $r(n)$ steps,
- β_2 can be reached from β_1 in less than $r(n)$ steps.

If there are two such configurations, M^D checks if the distance between $[r(n)]^4$ and $k - [r(n)]^4$ is equal to $i + q(j - i)$ for some natural q. In order to check if β_1 is accessible from β (or if β_2 is accessible from β_1) in less than $r(n)$ steps, M^D uses the divide-and-conquer algorithm of Savitch.   □

11. Space Hierarchy

In this chapter we show that the space hierarchy is dense, or in other words, that more languages can be accepted with more space. In Section 11.1 we use diagonalization to prove that for any fully space constructible function $L(n)$ there exists a language $A \subset \{1\}^*$, which is in $DSPACE[L(n)]$ or $(NSPACE[L(n)])$ but not in $DSPACE[L_1(n)]$ (or $NSPACE[L_1(n)]$) for any $L_1(n) = o(L(n))$. In Section 11.2 we present the proof of Hopcroft and Ullman (1967) that deterministic and nondeterministic space hierarchy is dense below logarithm.

11.1. Diagonalization

Stearns et. al. (1965) showed using diagonalization that the deterministic space hierarchy above logarithm is dense. They proved that if $L(n) \geq \log n$ and $L(n)$ is space constructible, then there exists a language accepted by an $L(n)$ space-bounded deterministic Turing machine and not accepted by any $Q(n)$ space-bounded deterministic Turing machine if $\liminf \frac{Q(n)}{L(n)} = 0$. Sipser (1980) showed that the diagonalization method can also be used for deterministic space below logarithm, and Immerman-Szelepcsényi theorem (Immerman 1988; Szelepcsényi 1988; cf. Thm. 7.3.1) allows us to use the diagonalization for nondeterministic space above logarithm. To illustrate the diagonalization method we prove the following theorem.

Theorem 11.1.1. Let $L_2(n)$ be a fully space constructible function, Then there exists a language A over one-letter alphabet, $A \subset \{1\}^*$, which is accepted by some $L_2(n)$ space-bounded deterministic (or nondeterministic) Turing machine M_s and not accepted by any strongly $L_1(n)$ space-bounded deterministic (or nondeterministic, respectively) Turing machine if $L_1(n) = o(L_2(n))$.

Proof. We shall prove the theorem for nondeterministic Turing machines. The proof for deterministic Turing machine is similar. The language A is constructed by diagonalization. Consider an enumeration of nondeterministic Turing machines with input alphabet $\{1\}$. We construct a nondeterministic Turing machine M which uses $L_2(n)$ space and disagrees on at least one input with any strongly $L_1(n)$ space-bounded nondeterministic Turing machine.

On an input 1^n, M first marks off $L_2(n)$ cells on the work tape (this is possible, since $L_2(n)$ is fully space constructible). M will not use more space than the marked $L_2(n)$. Next, M decomposes n into the form $n = 2^k m$ with odd m, and then simulates on the input 1^n the Turing machine M_k, i.e., the Turing machine with code number k. The machine M accepts if M_k reaches an rejecting state. Decomposition of n into the form $n = 2^k m$ requires at most k space and if M_k is strongly $L_1(n)$ space-bounded then the simulation requires $\log t \cdot L_1(n)$ space, where t is the number of work tape symbols of M_k.

Let us now consider the language A accepted by M, $A = L(M)$. Of course

$$A \in NSPACE(L_2(n)).$$

But

$$A \notin strong\text{-}NSPACE(L_1(n)).$$

Indeed, suppose that there is a strongly $L_1(n)$ space-bounded nondeterministic Turing machine accepting A. By Theorem 7.4.1, A is also accepted by a strongly $L_1(n)$ space-bounded recognizing Turing machine M_k with t work tape symbols. Since $L_1(n) = o(L_2(n))$, there exists $n = 2^k m$ for some sufficiently large odd m such that

$$\log t \cdot L_1(n) + k < L_2(n).$$

Hence, on input 1^n, M has sufficient space to decompose n and to simulate any computation of M_k, and M accepts 1^n if and only if M_k rejects, a contradiction. □

11.2. Space Hierarchy below Logarithm

Note that Theorem 11.1.1 is of little use below logarithm, because, as we showed in Section 6.2, no unbounded non-decreasing $o(\log n)$ function is fully space constructible. In this Section we present the result of (Hopcroft and Ullman 1967) who proved that if $L(n)$ is constructible and $L(n) \leq \log(n)$ then there exists a language A accepted by a strongly $L(n)$ space-bounded deterministic Turing machine, but not accepted by any weakly $Q(n)$ space-bounded nondeterministic Turing machine if $\liminf \frac{Q(n)}{L(n)} = 0$. Ito et. al. (1987) showed that also the alternating space hierarchy is dense by showing that the language A described in (Hopcroft and Ullman 1967) cannot be accepted by any $Q(n)$ space-bounded alternating Turing machine.

Let us recall that by Lemma 6.1.4(d), any function $K(\log \log n)$ is constructible if $K(n)$ is fully space constructible and $K(n) \geq n$. In particular, any function $[\log \log n]^r$ with rational $r \geq 1$, is space constructible.

First we present the proof of Hopcroft and Ullman (1967).

Theorem 11.2.1.   If $L(n)$ is constructible and $L(n) \leq \log(n)$, then there exists a language A accepted by a strongly $L(n)$ space-bounded deterministic Turing machine, but by no weakly $Q(n)$ space-bounded nondeterministic Turing machine if $\liminf \frac{Q(n)}{L(n)} = 0$.

Proof.   Let M_c be a Turing machine that constructs $L(n)$ and let Σ be the input alphabet of M_c. Let $\Sigma' = \Sigma \times \{a, b, c\}$. For every $(x, y) \in \Sigma'$, let $h_1((x, y)) = x$

and $h_2((x, y)) = y$. Extend the homomorphisms h_1 and h_2 to words over Σ' in the obvious way. Let A be the language consisting of words w over Σ' having the following property: If w is of length n and $h_1(w)$ causes M_c to use k space, then $h_2(w)$ is of the form $uc^m u$ for some word $u \in \{a, b\}^*$ of length $|u| = 2^k$, and $m = n - 2^{k+1}$.

It is easy to see that A is accepted by a strongly $L(n)$ space-bounded deterministic Turing machine. The machine first lays off $k \leq L(n)$ cells of work tape by simulating M_c on $h_1(w)$, then checks if $h_2(w)$ is of the form $h_2(w) = uc^m v$ with $|u| = |v| = 2^k$, and finally compares u and v symbol by symbol, using its work tape to keep the current position within u or v.

Suppose now, that A is accepted by a weakly $Q(n)$ space-bounded nondeterministic Turing machine M. For each n, let y be a word of length n over the alphabet Σ such that M_c uses exactly $L(n)$ space when processing y. Such a word y exists because $L(n)$ is space constructible. Consider all w in A such that $h_1(w) = y$. Then $h_2(w) = uc^m u$, where $u \in \{a, b\}^*$ and $|u| = 2^{L(n)}$. Observe now that if there are two words $w_1 = p_1 q_1$ and $w_2 = p_2 q_2$ of length n in A such that

- $h_1(w_1) = h_1(w_2) = y$,
- $|q_1| = |q_2| = 2^{L(n)}$, and
- $h_2(q_1) \neq h_2(q_2)$,

then q_1 and q_2 must belong to different equivalence classes of the relation $\equiv_{Q(n)}$ (defined in Definition 5.1.1), since otherwise, by Lemma 5.1.2, M accepts $p_1 q_2$ which is not in A.

According to Lemma 2.3.2 the number of $Q(n)$ space-bounded internal configurations of M is bounded by $d^{Q(n)}$ for some constant d. And there are at most

$$4^{(d^{Q(n)})^2}$$

possible equivalence classes of the relation $\equiv_{Q(n)}$. On the other hand, there are $2^{2^{L(n)}}$ different words $w = pq \in A$ with $h_1(w) = y$ and $|q| = 2^{L(n)}$. Thus,

$$2^{2^{L(n)}} \leq 4^{d^{2Q(n)}}.$$

Taking logarithm twice we obtain

$$L(n) \leq d_1 Q(n)$$

for some constant d_1. But then $\liminf \frac{Q(n)}{L(n)} = 0$ could not be true.   □

Ito et. al. (1987) proved that the language A described in the proof of Theorem 11.2.1 cannot be accepted by any $Q(n)$ space-bounded alternating Turing machine.

Theorem 11.2.2. If $L(n)$ is constructible and $L(n) \leq \log(n)$, then there exists a language A accepted by a strongly $L(n)$ space-bounded deterministic Turing machine, but by no weakly $Q(n)$ space-bounded alternating Turing machine if $\liminf \frac{Q(n)}{L(n)} = 0$.

Proof. Consider the language A defined in the proof of Theorem 11.2.1 Suppose that A is accepted by a weakly $Q(n)$ space-bounded alternating Turing machine M. For each n, let y be a word of length n over the alphabet Σ such that M_c uses exactly $L(n)$ space when processing y. Consider all w in A such that $h_1(w) = y$. Then $h_2(w) = uc^m u$, where $u \in \{a, b\}^*$ and $|u| = 2^{L(n)}$. For each input $w \in A$, choose one $Q(n)$ space-bounded accepting computation tree. We shall denote this tree by $T(w)$. Consider now two words $w_1 = p_1 q_1$ and $w_2 = p_2 q_2$ of length n in A such that $h_1(w_1) = h_1(w_2) = y$ and $|q_1| = |q_2| = 2^{L(n)}$. We say that w_1 and w_2 are equivalent, denoted $w_1 \cong_{Q(n)} w_2$ if

- for every two $Q(n)$ space-bounded internal configurations β_1, β_2,
      there is a computation path of $T(w_1)$ that starting in β_1 at the first letter
      of q_1 reaches afterwards β_2 just after it leaves q_1,
  if and only if
      there is a computation path of $T(w_2)$ that starting in β_1 at the first letter
      of q_2 reaches afterwards β_2 just after it leaves q_2, and
- for every $Q(n)$ space-bounded internal configuration, β
      there is a computation path of $T(w_1)$ that starting in β at the first letter
      of q_1 accepts without leaving q_1.
  if and only if
      there is a computation path of $T(w_2)$ that starting in β at the first letter
      of q_2 accepts without leaving q_2.

Note that
$$\text{if}\quad h_2(q_1) \neq h_2(q_2)\quad \text{then}\quad w_1 \not\cong_{Q(n)} w_2,$$
since otherwise, similarly as in Lemma 5.1.2, one could show that M accepts $p_1 q_2$ which is not in A, a contradiction.

There are at most
$$4^{d^2 Q(n)}$$
possible equivalence classes of the relation $\cong_{Q(n)}$. On the other hand, there are $2^{2^{L(n)}}$ words $w = pq$ with different $h_2(q)$ and $|q| = 2^{L(n)}$. Thus, similarly to the proof of Theorem 11.2.1 one can derive $L(n) \leq dQ(n)$, a contradiction. $\square$

12. Closure under Concatenation

In this chapter we discuss closure under concatenation of the space complexity classes. It is easy to see that if the function $L(n)$ is monotone and $L(n) \geq \log n$, then the classes $DSPACE[L(n)]$ and $NSPACE[L(n)]$ are closed under concatenation. Using $\log n$ space the Turing machine can remember the position of any point on an input. So, it can keep in its memory the position where two subwords u and v are concatenated, and check if u belongs to one of the languages and v to the other.

This is not the case for sublogarithmic space. Ibarra and Ravikumar (1988) presented a proof that $strong\text{-}DSPACE[L(n)]$ is not closed under concatenation if $L(n) \geq \log \log n$ and $L(n) = o(\log n)$. We shall show that then also $strong\text{-}NSPACE[L(n)]$ is not closed under concatenation, and correct some minor errors of (Ibarra and Ravikumar 1988).

Consider the language

$$A = \{a^r b^t \mid F(r) = F(s) \text{ for some } s \leq t\},$$

where

$$F(n) = \min\{i \mid i \text{ does not divide } n\}.$$

Then we have

Lemma 12.1. If $L(n) = o(\log n)$, then

$$A \notin strong\text{-}NSPACE[L(n)].$$

Proof. Suppose for a contradiction that

$$A \in strong\text{-}NSPACE[L(n)].$$

Then by Theorem 7.4.2, also the complement of A

$$A^c \in strong\text{-}NSPACE[L(n)].$$

Furthermore, the class $strong\text{-}NSPACE[L(n)]$ is closed under intersection with regular languages. Hence, if

$$B = A^c \cap a^* b^* = \{a^r b^t \mid F(r) \neq F(s) \text{ for any } s \leq t\},$$

then

$$B \in strong\text{-}NSPACE[L(n)].$$

Observe now the following:

(1) for any i and j, $a^i b^j \in B$ implies $i > j$, and
(2) there are infinitely many m such that $a^m b^{m-1} \in B$.

(1) follows immediately from the definition of B.
To prove (2) let us consider

$$m = lcm\{i \mid i \le r\} \qquad \text{for some } r,$$

where lcm denotes the least common multiple. Then by the definition of $F(m)$,

$$F(m) > r. \tag{3}$$

On the other hand,

$$F(k) \le r \qquad \text{for each } k < m, \tag{4}$$

since, otherwise, every $i \le r$ divides k, so k is a common multiple of all $i \le r$ less than m, a contradiction. Hence, by (3) and (4), $F(m) \ne F(k)$ for all $k \le m - 1$, and

$$a^m b^{m-1} \in B.$$

And there is infinitely many m of the form $m = lcm\{i \mid i < r\}$ for some r, so

$$a^m b^{m-1} \in B \qquad \text{for infinitely many } m.$$

Similarly as in the proof of Theorem 6.2.4 one can show that since $L(n) = o(\log n)$, there is an input $w = a^m b^{m-1} \in B$ such that the number of all possible $L(|w|)$ space-bounded internal configurations of M is less than $m - 1$, and also $a^m b^{m-1+m!}$ is accepted by M, but this contradicts (1). Hence, B cannot be accepted by any strongly $L(n)$ space-bounded nondeterministic Turing machine. $\square$

Now we are ready to formulate the main theorem of this chapter.

Theorem 12.2. If $L(n) \ge \log\log n$ and $L(n) = o(\log n)$, then neither *strong-NSPACE*$[L(n)]$ nor *strong-DSPACE*$[L(n)]$ is closed under concatenation.

Proof. Consider the language

$$C = \{a^r b^s \mid F(r) = F(s)\}.$$

It is easy to see that the language C can be accepted by a strongly $\log\log n$ space-bounded deterministic Turing machine. The Turing machine computes and compares $F(r)$ and $F(s)$ using $O(\log\log(r + s))$ space (cf. Section 3.1).

On the other hand, note that the concatenation of C and the regular language b^* is equal to

$$A = Cb^* = \{a^r b^t \mid F(r) = F(s) \text{ for some } s \le t\}$$

and by Lemma 12.1,

$$A \notin \text{strong-}NSPACE[L(n)].$$

Thus, C and b^* are in *strong-DSPACE*$[L(n)]$ and in *strong-NSPACE*$[L(n)]$, but their concatenation Cb^* is neither in *strong-NSPACE*$[L(n)]$ nor in *strong-DSPACE*$[L(n)]$. $\square$

Note that, contrary to Lemma 2 of (Ibarra and Ravikumar 1988), there is a language

$$D = \{0^n 1^{F(n)} \mid n \geq 1\} \subseteq 0^* 1^*$$

such that

- D is strongly accepted in $\log \log n$ space,
- there are infinitely many m such that $0^{i_m} 1^m \in D$ for some i_m, and
- there is no m and i_m such that $0^{i_m} 1^{m+\lambda m!} \in D$ for every λ.

Note also that, contrary to an observation of (Ibarra and Ravikumar 1988), the concatenation of infinite regular tally (over a one-letter alphabet) language with another infinite (non-regular) language need not to be regular.

Consider two languages:

$$R = (aa)^* \cup a \qquad \text{and} \qquad E = \{a^{2^n} \mid n \geq 1\}$$

Then R is regular, but RE not. Indeed,

$$RE = \{a^{2^n+1} \mid n \geq 1\} \cup \{a^{2n} \mid n \geq 2\}$$

and the intersection of RE with the regular language $a(aa)^*$:

$$RE \cap a(aa)^* = \{a^{2^n+1} \mid n \geq 1\}$$

is not regular, hence RE is not regular.

13. Alternating Hierarchy

In this chapter we present a proof that the alternating hierarchy for $SPACE$ $[L(n)]$ collapses to Σ_1 level if $L(n) \geq \log n$, and in contrast to this, we show that the alternating hierarchy does not collapse to the Σ_1 level if $L(n) = o(\log n)$.

Immerman (1988) and Szelepcsényi (1988) (cf. Thm. 7.3.2) proved that if $L(n) \geq \log n$, then the nondeterministic $L(n)$ space complexity class is closed under complement. This implies that the strongly $L(n)$ space-bounded alternating hierarchy collapses to $NSPACE[L(n)]$.

Below logarithm the situation is quite different. There exists a language over a one-letter alphabet that can be accepted by a $\log \log n$ space-bounded Π_2-alternating Turing machine but by no $o(\log n)$ space-bounded nondeterministic Turing machine. The first proof of this fact was presented by Chang et. al. (1987). They proved that below $\log n$ alternating Turing machines accept more than nondeterministic ones and although they did not say that explicitly, their alternating Turing machine was actually Π_2-alternating. In (Szepietowski 1989) a slightly stronger result was presented. It was shown that also for any unbounded fully space constructible function $S(n) = o(\log n)$ there exists a language over a one-letter alphabet that can be accepted by an $S(n)$ space-bounded Π_2-alternating Turing machine but not by any $o(\log n)$ space-bounded nondeterministic Turing machine. This means that if $L(n) = o(\log n)$ and $L(n) \geq S(n)$ for some unbounded fully space constructible $S(n)$, then

$$NSPACE[L(n)] \subsetneqq \Pi_2\text{-}SPACE[L(n)]. \tag{1}$$

Hence, the $L(n)$ space-bounded alternating hierarchy does not collapse to Σ_1 level.

Recall that there exist unbounded fully space constructible functions $S(n)$ such that $S(n) \leq \log \log n$, so (1) is also valid for $L(n) \geq \log \log n$ and $L(n) = o(\log n)$.

Recently von Braunmühr, Gengler, and Rettinger (1994) and Liśkiewicz and Reischuk (1994) proved that the alternating hierarchy for $SPACE[L(n)]$ is infinite if $L(n) \geq \log \log n$ and $L(n) = o(\log n)$.

13.1. Collapsing of the Alternating Hierarchy above Logarithm

In this section we present a proof that the strongly $L(n)$ space-bounded alternating hierarchy collapses to Σ_1 level if $L(n) \geq \log n$ (Immerman 1988)). Recall (cf. Theorem 8.1.1) that it does not matter which mode of space complexity is used if $L(n)$ is fully space constructible, and this is the case for all "normal" functions $L(n) \geq \log n$.

Theorem 13.1.1. If $L(n) \geq \log n$, then for every natural k,

$$strong\text{-}\Pi_k\text{-}SPACE[L(n)] = strong\text{-}\Sigma_1\text{-}SPACE[L(n)] \quad \text{and}$$
$$strong\text{-}\Sigma_k\text{-}SPACE[L(n)] = strong\text{-}\Sigma_1\text{-}SPACE[L(n)].$$

Recall that $\Sigma_1\text{-}SPACE[L(n)] = NSPACE[L(n)]$.

Proof. Collapsing of the $L(n)$ space-bounded alternating hierarchy follows from two facts. First, that the class of strongly $L(n)$ space-bounded nondeterministic Turing machines is closed under so-called independent (of the starting configuration) complement. In other words, every strongly $L(n)$ space-bounded Turing machine M has its independent complement M^c such that, M^c is strongly $L(n)$ space-bounded, and for each input x and each $L(|x|)$ bounded configuration β of M, M^c starting on x in β accepts if and only if M starting on x in β does not accept (the set of states of M^c contains the states of M and the set of work tape symbols of M^c contains the work tape symbols of M). M^c remembers the configuration β and counts all configurations reachable from β, by using the method of Immerman and Szelepcsényi (see Theorem 7.3.1).

The second fact is that since $L(n) \geq \log n$, it is possible to detect loops (cf. Section 7.1) and to cut off infinite computations. Thus, we can assume that every computation is finite and ends either with an accepting or rejecting state.

Suppose now that M^c is an independent complement of M and β is an arbitrary accessible configuration of M. Let T be the finite tree labeled by all configurations accessible by M from β and let T^c be the tree labeled by all configurations accessible from β by M^c. T^c can be made finite, since we can detect loops. Note that if there is an accepting leaf in T, then all leaves in T^c are rejecting, and vice versa, if all leaves in T are rejecting, then there is an accepting leaf in T^c.

Now we can show that if $k \geq 1$, then for every Π_{k+1}- or Σ_{k+1}-alternating Turing machine M_1, there is a Π_k- or Σ_k- (respectively) alternating Turing machine M_2 that accepts the same language. Suppose that M_1 is Σ_{k+1}-alternating. Firstly, M_2 simulates M_1 as long as it reaches the kth alternation for the first time, say in a configuration β. Let T be a finite tree of all configurations accessible by M_1 from β. Note that T consists entirely of existential configurations if k is even, and entirely of universal configurations if k is odd. In M_2 the tree T will be exchanged for a finite tree T^c in such a way that:

- T^c is entirely universal if T is existential (for even k), and T^c is entirely existential if T is universal (for odd k),
- T^c is rooted at β' which is equal to β except that β' is existential if β is universal, and vice versa, and
- β leads M_1 to acceptance if and only if β' leads M_2 to acceptance.

Thus, M_2 is k alternation-bounded and accepts the same language as M_1. What remains, is to show how M_2 exchanges the tree T for T^c.

Suppose first that k is even and β (and entire T) is existential. From the configuration β, M_2 simulates the independent complement of M_1, say M_3. More precisely, M_3 is an independent complement of the nondeterministic Turing machine which works like M_1 but all its states are existential. At the end, M_2 accepts if M_3 rejects, and vice versa, M_2 rejects if M_3 accepts. The configuration β is exchanged for the configuration β' that is equal to β except that it is universal and all configurations accessible from β' are universal in M_2.

Note that if β leads M_1 to acceptance, then there is an accepting configuration accessible by M_1 from β, so all computations of M_3 which are started in β lead to rejection. Hence, all computations of M_2 which are started in β' lead to the accepting state. Thus, β' leads M_2 to acceptance.

On the other hand, if β leads M_1 to rejection, then all computations of M_1 which are started in β lead to rejection. Hence, there is a computation that leads M_3 to acceptance and there is a computation of M_2 that starting in β' leads to rejection. Thus, β' leads M_2 to rejection.

If k is odd and β (and entire T) is universal, then from the configuration β, M_2 simulates the independent complement of the nondeterministic Turing machine M_4 such that: M_4 works exactly like M_1 but all its states are existential and rejecting states in M_1 are accepting in M_4, and vice versa, accepting states in M_1 are rejecting in M_4. The configuration β is exchanged for the configuration β' that is existential and all configurations accessible from β' are existential in M_2. Similarly as above one can show that β leads M_1 to acceptance if and only if β' leads M_2 to acceptance.

Note that if $k = 0$, then β is the initial configuration, M_1 is Σ_1-alternating or nondeterministic, and M_2 is Π_1-alternating.

Thus, we have shown how to reduce every Σ_{k+1}-alternating Turing machine to a Σ_k-alternating one. One needs to apply this reduction $k-1$ times to reduce every Σ_k-alternating Turing machine to a Σ_1-alternating one. Hence, for every natural $k \geq 1$,

$$\Sigma_k\text{-}SPACE[L(n)] \subseteq \Sigma_1\text{-}SPACE[L(n)]$$

and

$$\Sigma_1\text{-}SPACE[L(n)] \subseteq \Pi_1\text{-}SPACE[L(n)].$$

Similarly, one can show that for every $k \geq 1$,

$$\Pi_k\text{-}SPACE[L(n)] \subseteq \Pi_1\text{-}SPACE[L(n)]$$

and

$$\Pi_1\text{-}SPACE[L(n)] \subseteq \Sigma_1\text{-}SPACE[L(n)].$$

On the other hand, for every natural $k > 1$, we have the trivial inclusions

$$\Sigma_1\text{-}SPACE[L(n)] \subseteq \Sigma_k\text{-}SPACE[L(n)] \subseteq \Pi_{k+1}\text{-}SPACE[L(n)].$$

Thus, we have proven Theorem 13.1.1.   □

13.2. Noncollapsing of the Alternating Hierarchy below Logarithm

In this section we present the proof of (Szepietowski 1989) that below logarithm the alternating hierarchy does not collapse to Σ_1 level. Recently von Braunmuhr, Gengler, and Rettinger (1994) and Liśkiewicz and Reischuk (1994) proved that the alternating hierarchy for $SPACE[L(n)]$ is infinite if $L(n) \geq \log\log n$ and $L(n) = o(\log n)$.

Theorem 13.2.1. Let $S(n)$ be an unbounded fully space constructible function, $S(n) = o(\log n)$, and $L(n) = o(\log n)$. Then there exists a language $A \subseteq a^*$ such that:

- $A \in \Pi_2\text{-}SPACE[S(n)]$ and
- $A \notin weak\text{-}NSPACE[L(n)]$.

As an immediate corollary we have the following theorem.

Theorem 13.2.2. Let $L(n)$ be a function, $L(n) = o(\log n)$, and $L(n) \geq S(n)$ for some unbounded fully space constructible function $S(n)$. Then

$$NSPACE[L(n)] \subsetneq \Pi_2\text{-}SPACE[L(n)].$$

The theorem is valid for both the strong and weak mode of space complexity.

Recall that there exist unbounded fully space constructible functions $S(n)$ such that $S(n) \leq \log\log n$, e.g. $S(n) = \log[\min\{i \mid i \text{ does not divide } n\}]$ (cf. Chapter 6). So, Theorem 13.2.1 is valid if $L(n) \geq \log\log n$ and $L(n) = o(\log n)$.

Observe that, by Theorem 13.1.1, $\Pi_k\text{-}SPACE[L(n)]$ and $\Sigma_k\text{-}SPACE[L(n)]$ are both contained in $NSPACE[\log n]$ for any $k \geq 1$ and $L(n) \leq \log n$.

Proof of Theorem 13.2.1 For any natural d, consider the language

$$A_d = \{a^n \mid n \leq lcm\{i \mid i \leq 2^{d \cdot S(n)}\}\},$$

where $lcm\,E$ denotes the least common multiple of the elements of E.
    First we shall show that for any natural d,

$$A_d \in \Pi_k\text{-}SPACE[L(n)]$$

The $S(n)$ space-bounded Π_2-alternating Turing machine M accepting A_d works as follows. On an input of length n, M first deterministically constructs the function $S(n)$. M will not use more space than $S(n)$. Then it checks if $n \leq lcm\{i \mid i \leq 2^{d \cdot S(n)}\}$, by checking if for all $k < n$, there exists $i \leq 2^{d \cdot S(n)}$ such that i does not divide k. This can be done as follows. Branching universally, M moves to the k-th cell of the input, then (existentially) guesses $i \leq 2^{d \cdot S(n)}$ and going from the k-th to the first cell checks if i divides k, and accepts if this is not the case. Note that M has to store i on its work tape, but not k.

Now we shall prove that for some d, A_d is not in $weak\text{-}NSPACE[o(\log n)]$. To do this, we need the following lemma.

Lemma 13.2.3.   There exists a constant $d > 0$ such that:

(a) A_d is infinite and
(b) $A_d \subseteq \{a^n \mid S(n) \geq c \cdot \log\log n\}$ for some constant $c > 0$.

Proof of Lemma 13.2.3 (a) We show that if the constant d is large enough, then A_d is infinite. Since $S(n)$ is unbounded, by Theorem 5.1.3, there is a constant $d_1 > 0$ such that

$$S(n) \geq d_1 \cdot \log\log n \qquad \text{for infinitely many } n.$$

By Lemma 4.1.2(b), there is a constant $c_1 > 1$ such that

$$c_1^{2^{d \cdot S(n)}} \leq lcm\{i \mid i \leq 2^{d \cdot S(n)}\}.$$

Choose d in such a way that

$$\left[S(n) \geq d_1 \cdot \log\log n \right] \implies \left[n \leq c_1^{2^{d \cdot S(n)}} \right].$$

It is obvious that such a constant d exists. Thus,

$$\left[S(n) \geq d_1 \cdot \log\log n \right] \implies \left[n \leq c_1^{2^{d \cdot S(n)}} \right] \implies$$

$$\implies \left[n \leq lcm\{i \mid i \leq 2^{d \cdot S(n)}\} \right] \implies \left[a^n \in A_d \right],$$

hence,

$$a^n \in A_d \qquad \text{for infinitely many } n.$$

(b) Let d be an arbitrary constant, $d > 0$. By Lemma 4.1.2(b),

$$lcm\{i \mid i \leq 2^{d \cdot S(n)}\} \leq c_2^{2^{d \cdot S(n)}},$$

so there is a constant $c > 0$ such that

$$\left[a^n \in A_d \right] \implies \left[n \leq lcm\{i \mid i \leq 2^{d \cdot S(n)}\} \right] \implies$$

$$\implies \left[n \leq c_2^{2^{d \cdot S(n)}} \right] \implies \left[S(n) \geq c \cdot \log\log n \right].$$

Thus, we have shown that for any $d > 0$, there exists a constant $c > 0$ such that

$$A_d \subseteq \{a^n \mid S(n) \geq c \cdot \log\log n\}.$$

This completes the proof of Lemma 13.2.3.   □

Now we shall show that if d is as described in Lemma 13.2.3 and $L(n) = o(\log n)$, then

$$A_d \notin weak\text{-}NSPACE[L(n)].$$

Suppose, for a contradiction, that

$$A_d \in weak\text{-}NSPACE[L(n)],$$

then by Theorem 6.2.4, there is an infinite regular language R such that $R \subseteq A_d$ and by Lemma 13.2.3(b), also

$$R \subseteq \{a^n \mid S(n) \geq c \cdot \log\log n\},$$

but this contradicts Theorem 6.2.9.   □

14. Independent Complement

In Chapter 13 we showed that the alternating hierarchy for $SPACE[L(n)]$ collapses to Σ_1 level if $L(n) \geq \log n$ and it does not collapse if $L(n) = o(\log n)$. Taking these two facts into consideration, one can raise a natural question: does this mean that the nondeterministic sublogarithmic space complexity classes are not closed under complement? We discuss this problem in this chapter and although we do not provide a complete answer, we prove some weaker result. But first we must formulate two definitions.

Definition 14.1. *We say that a strongly $L(n)$ space-bounded Turing machine $M_c = \langle Q_c, \Sigma, \Gamma_c, \delta_c, q_{0c}, F_c \rangle$ is an independent (of the starting configuration) complement of another strongly $L(n)$ space-bounded Turing machine $M = \langle Q, \Sigma, \Gamma, \delta, q_0, F \rangle$ if*

- *$Q \subseteq Q_c$, $\Gamma \subseteq \Gamma_c$, and*
- *for each input $x \in \Sigma^*$ and each $L(|x|)$ space-bounded configuration β of M, M_c starting on x in β accepts if and only if M starting on x in β does not accept.*

Definition 14.2. *We say that a strongly $L(n)$ space-bounded Turing machine $M_s = \langle Q_s, \Sigma, \Gamma_s, \delta_s, q_{0s}, F_s \rangle$ simulates independently (of the starting configuration) a strongly $L(n)$ space-bounded Turing machine $M = \langle Q, \Sigma, \Gamma, \delta, q_0, F \rangle$ if*

- *$Q \subseteq Q_s$, $\Gamma \subseteq \Gamma_s$, and*
- *for each input $x \in \Sigma^*$ and each $L(|x|)$ space-bounded configuration β of M, M_s starting on x in β accepts if and only if M starting on x in β accepts.*

If $L(n) \geq \log n$, then, as we observed in the proof of Theorem 13.1.1, the class of strongly $L(n)$ space-bounded nondeterministic Turing machines is closed under independent complement. It is possible then to remember the configuration β and to count all configurations reachable from β, by using the method of Immerman (1988) and Szelepcsényi (1988) (cf. Theorem 7.3.1). It is this fact and the ability of such Turing machines to detect loops in its behavior that is needed to prove that the alternating hierarchy collapses.

In this Chapter we show that strongly $o(\log n)$ space-bounded nondeterministic Turing machines are not closed under independent complement, even if the input alphabet consists of one letter only. This contrast with the fact that the class of tally languages accepted by strongly $L(n)$ space-bounded nondeterministic Turing machines is closed under complement for arbitrary function $L(n)$ (Th. 7.4.2).

On the other hand, as we showed in Theorem 7.2.4, the class of strongly $L(n)$ space-bounded deterministic Turing machines with a one-letter input alphabet is closed under independent complement, for arbitrary $L(n)$. From that fact it follows that there exists an $L(n)$ space-bounded nondeterministic Turing machine (even with a one-letter input alphabet) which cannot be simulated independently by any $L(n)$ space-bounded deterministic Turing machine.

It is unknown whether strongly $L(n)$ space-bounded deterministic Turing machines with a multi-letter input alphabet are closed under independent complement. It seems that proof of (Sipser 1980; cf. Theorem 7.2.1) cannot be applied to independent complement.

A very similar notion of independent simulation was introduced by Kannan (1983). Using similar arguments (but not discussing detecting of loops) he sketched a proof that for any function $L(n)$ such that $L(n) \geq \log\log n$ and $L(n) = o(\log n)$, there is an $L(n)$ space-bounded nondeterministic Turing machine with a multi-letter input alphabet which cannot be "simulated" by any $L(n)$ space-bounded deterministic Turing machine. Independent simulation of finite automata was also discussed by Birget (1989).

Recall (Chapter 7) that if $L(n) \geq \log\log n$ and $L(n) = o(\log n)$, then the class of languages accepted by weakly $L(n)$ space-bounded Turing machines (deterministic or nondeterministic) is not closed under complement. Hence, weakly space-bounded Turing machines are not closed under complement.

Now we present the main Theorem of this chapter.

Theorem 14.3. If $L(n) = o(\log n)$ and $L(n) \geq S(n)$ for some unbounded fully space constructible $S(n)$, then the class of strongly $L(n)$ space-bounded nondeterministic Turing machines (even with a one-letter input alphabet) is not closed under independent complement.

Proof. Suppose, for a contradiction, that every strongly $L(n)$ space-bounded nondeterministic Turing machine with a one-letter input alphabet has its independent complement. Let $S(n)$ be an unbounded fully space constructible function such that $S(n) \leq L(n)$. By Theorem 13.2.1, there exists a language A which belongs to $\Pi_2\text{-}SPACE[S(n)]$ and not to $NSPACE[L(n)]$. Let M be the strongly $S(n)$ space-bounded Π_2-alternating Turing machine accepting A described in the proof of Theorem 13.2.1. We can assume that each computation of M is finite and ends either with an accepting state s_a or a rejecting state s_r.

Consider now two Turing machines:

- M_1 which is strongly $S(n)$ space-bounded Σ_2-alternating and works exactly like M, but existential states in M are universal in M_1, and vice versa, universal states in M are existential in M_1; and the state s_r is accepting in M_1 and s_a is rejecting. Note that M_1 accepts A^c –the complement of A.

- M_2 which is strongly $L(n)$ space-bounded nondeterministic and is an independent complement of M, or more precisely, of the nondeterministic Turing machine which works exactly like M, but has only existential states.

Similarly as in the proof of Theorem 13.1.1 we shall reduce M_1 to a strongly $L(n)$ space-bounded nondeterministic Turing machine M_3 which also accepts A^c. But now we do not need to detect loops and cut off infinite computations.

First, M_3 acts like M_1 until for the first time it reaches a universal configuration, say β. Then it simulates M_2 starting from β. All configurations of M_3 are existential.

Note that if β leads M_1 to acceptance, then all computations of M_1 (and of M) starting in β are finite and end with s_r (s_r is accepting in M_1 and rejecting in M), so there exists an accepting computation of M_2 starting in β. Hence, β leads M_3 to acceptance.

On the other hand, if β leads M_1 to rejection, then there exists a computation of M_1 (and of M) starting in β that ends with s_a (s_a is rejecting in M_1 and accepting in M, and all computations of M_1 are finite). So, there is no computation in M_2 that leads to acceptance. Hence, β does not lead M_3 to acceptance.

Thus, M_1 and M_3 accept the same language A^c. So,

$$A^c \in strong\text{-}NSPACE[L(n)],$$

but by our assumption, $strong\text{-}NSPACE[L(n)]$ is closed under complement, so also

$$A \in strong\text{-}NSPACE[L(n)],$$

but this contradicts Theorem 13.2.1. □

On the other hand, as we showed in Theorem 7.2.4, the class of strongly $L(n)$ space-bounded deterministic Turing machines with a one-letter input alphabet is closed under independent complement, for arbitrary $L(n)$. Hence, we have the following theorem.

Theorem 14.4. If $L(n) = o(\log n)$ and $L(n) \geq S(n)$ for some unbounded fully space constructible $S(n)$, then there is a strongly $L(n)$ space-bounded nondeterministic Turing machine (even with a one-letter input alphabet) which cannot be independently simulated by any strongly $L(n)$ space-bounded deterministic Turing machine.

15. Other Models of Turing Machines

In this chapter we consider some other models of Turing machines with bounded space and see which functions are space constructible by these machines.

We consider

- two-dimensional Turing machines,
- 1-inkdot Turing machines,
- 1-pebble Turing machines, and
- demon Turing machines.

An input tape of a two-dimensional Turing machine is a two-dimensional rectangular array of symbols from some alphabet Σ. A 1-inkdot Turing machine is a usual Turing machine with the additional power of marking one cell on the input tape (with an inkdot). This cell is marked once and for all (no erasing) and no more than one dot of ink is available. A 1-pebble Turing machine is equipped with a pebble which can be placed on and removed from any cell of the input tape. A demon $L(n)$ Turing machine is a Turing machine that has $L(n)$ space marked off on its work tape for every input of length n, and cannot use more space than $L(n)$.

These models do not differ much from the normal Turing machine if the space bound function $L(n) \geq \log n$, but they behave differently if $L(n) = o(\log n)$. This shows that Turing machines with sublogarithmic space are sensitive to modifications of the definition.

In (Szepietowski 1992) it was shown that the function $\log \log n$ is fully space constructible and functions: $\log^{(k)} n$ for every k, and $\log^* n = \min\{i \mid \log^{(i)} n \leq 1\}$ are space constructible by two-dimensional Turing machines. Thus, the space hierarchy of two-dimensional languages starts much lower than in the one-dimensional case. Recall (Ch. 5) that space used by any one-dimensional Turing machine is either bounded by a constant or it reaches $O(\log \log n)$ infinitely often.

Hartmanis and Ranjan (1989) showed that the function $\log \log n$ can be fully space constructed by a nondeterministic 1-inkdot Turing machine, and that for 1-inkdot Turing machines nondeterministic sublogarithmic space is more powerful than deterministic one.

Chang et. al. (1986) proved that $\log \log n$ is the lower bound for accepting non-regular languages by 1-pebble (and 1-inkdot) Turing machines. They also showed that $\log \log n$ is fully space constructed by a deterministic 1-pebble Turing machine and that $L(n)$ space-bounded 1-pebble Turing machines are more powerful than those without pebbles if $L(n) \geq \log \log n$ and $L(n) = o(\log n)$.

The demon Turing machine was introduced by Hartmanis and Ranjan (1989) to emphasize the importance of the constructibility of tape bounds. They considered two languages $\{a^k b^k \mid k \geq 0\}$ and $\{w \# w \mid w \in \{a, b\}^*\}$ which require $O(\log n)$ space to be accepted by a standard deterministic or nondeterministic Turing machine. And they showed that $\{a^k b^k \mid k \geq 0\}$ is accessible in $O(\log \log n)$ space by a demon Turing machine, but $\{w \# w \mid w \in \{a, b\}^*\}$ requires $O(\log n)$ space. This very clearly shows that the lower bound $O(\log n)$, on the amount of space required to recognize $\{a^k b^k \mid k \geq 0\}$ is purely due to space constructibility reasons, whereas $\{w \# w \mid w \in \{a, b\}^*\}$ requires $O(\log n)$ space independently of any constructibility properties of the Turing machine.

15.1. Two-dimensional Turing Machines

In this section we consider functions that are constructed by *two-dimensional Turing machines* with square inputs. We shall show that the function $\log \log n$ is fully space constructible and functions $\log^* n$ and $\log^{(k)} n$ for every k, are space constructible by two-dimensional Turing machines.

Let Σ be a finite set of symbols. A *two-dimensional tape* over Σ is a two-dimensional rectangular array of symbols from Σ. The set of all two-dimensional tapes over Σ is denoted by $\Sigma^{(2)}$. For each x in $\Sigma^{(2)}$, $x(i, j)$ denotes the symbol in x standing in the row i and column j.

We shall consider two-dimensional Turing machines with a read-only input tape and a separate work tape. The input tape consists of a two-dimensional tape surrounded by an additional symbol $\#$ and the input head starting from the top-left corner of the input can move right, left, down, or up. The work tape of two-dimensional Turing machines is one-dimensional.

For Turing machines with square inputs, we define 2-space constructible and fully 2-space constructible functions.

Definition 15.1.1. *A function $L(n)$ on natural numbers is 2-space constructible if there exists a strongly $L(n)$ space-bounded two-dimensional deterministic Turing machine M such that for every n, there exists a square input with n rows and n columns on which M uses exactly $L(n)$ space. $L(n)$ is called fully 2-space constructible if M uses $L(n)$ space on each square with n rows and n columns.*

The following lemma presents basic properties of 2-space constructible and fully 2-space constructible functions.

Lemma 15.1.2.
(a) If $L(n)$ is fully 2-space constructible, then $L(n)$ is 2-space constructible,
(b) if $L(n)$ is 2-space constructible and $L(n) \geq n^2$, then $L(n)$ is fully 2-space constructible,
(c) if $L(n)$ and $K(n)$ are (fully) 2-space constructible, then $L(n) + K(n)$ and $L(n)K(n)$ are (fully) 2-space constructible.
(d) if $L(n)$ and $K(n)$ are (fully) 2-space constructible and $K(n) \geq n$, then the composition $K(L(n))$ is (fully) 2-space constructible (by (f) below, $K(n)$ is then fully space constructible),

(e) If $L(n)$ is (fully) space constructible, then $L(n)$ is (fully) 2-space constructible,

(f) if $L(n)$ is fully 2-space constructible and $L(n) \geq \log n$, then $L(n)$ is fully space constructible,

(g) if $L(n)$ is 2-space constructible and $L(n) \geq \log n$, then $L(\sqrt{n})$ is space constructible.

Proof. (a)–(e) follow immediately from the definitions.

(f) If $L(n)$ is constructed by a two-dimensional deterministic Turing machine M, it can also be constructed by a one-dimensional deterministic Turing machine M' which simulates M on empty tapes (consisting of one symbol). M' remembers only the position of the input head of M which requires $\log n$ space.

(g) If space-bound $L(n) \geq \log n$, then any two-dimensional $L(n)$ space-bounded Turing machine can be simulated by a one-dimensional $L(\sqrt{n})$ space-bounded Turing machine M' such that M accepts a square tape x if and only if M' accepts one-dimensional word consisting of rows of x standing one after one.   □

First we show that $\log\log n$ is fully 2-space constructible (Szepietowski 1992).

Theorem 15.1.3. The function $\log\log n$ is fully 2-space constructible.

Proof. We should show that there exists a two-dimensional deterministic Turing machine M which uses $\log\log n$ space on every square input x with n rows and n columns. M starts from the boundary symbol # standing in the cell $x(1,0)$ and goes two steps right and one step up to the boundary cell $x(0,2)$, then it goes to $x(4,0)$ by repeating two steps down followed by one step left until it reaches the boundary symbol #, and next it goes to $x(0,8)$, by repeating two steps right and one step up. M continues these walks through x until it reaches for the first time # on the right or bottom edge of x. M counts in binary how many times it reaches the upper or left edge of x (the first cell $x(1,0)$ is not counted). M visits on the upper edge of x the cells $x(0,2^i)$ with odd $i \leq \log n$, and on the left edge $x(2^i,0)$ with even $i \leq \log n$. So, the number of cells visited by M on the left or upper edge is equal to $\lfloor \log n \rfloor$. Thus, at the end M keeps on its storage tape $\lfloor \log n \rfloor$ in binary using $\log\log n$ space.   □

To prove that the functions $\log^* n$ and $\log^{(k)} n$ for every k, are 2-space constructible, we shall need the following lemma.

Lemma 15.1.4. There exists a language

$$LOG = \{x_m \mid m \geq 1\} \subseteq \{0,1\}^{(2)}$$

such that LOG is accepted by a two-dimensional finite automaton A (Turing machine using zero space) and for every m,

- x_m is a square with m rows and m columns, and
- in the last row of x_m there are 1's in the first $\lfloor \log m \rfloor$ cells and 0's in the others.

Proof. First we describe the language LOG and then the finite automaton A accepting it. The language LOG consists of elements x_m for every m. To simplify the proof, we consider only $m \geq 8$ (the others can be accepted separately).

Here we present x_{12} as an example. For clarity, the dot stands for 1 and $\bigcirc$, O, and 0 stand for 0 ($\bigcirc$ stands for 0 on the main diagonal, O stands for 0 above the main diagonal, and 0 stands for 0 below the main diagonal).

```
 O  0  0  0  0  0  0  0  0  0  0  0
 ·  O  0  0  0  ·  ·  0  0  ·  ·  0
 ·  ·  O  ·  0  ·  ·  0  0  ·  ·  0
 ·  ·  0  O  0  0  0  0  0  ·  ·  0
 ·  ·  ·  0  O  ·  ·  ·  0  ·  ·  0
 ·  ·  ·  0  0  O  ·  ·  0  ·  ·  0
 ·  ·  ·  0  0  0  O  ·  0  ·  ·  0
 ·  ·  ·  0  0  0  0  O  ·  0  ·  ·  0
 ·  ·  ·  0  0  0  0  0  O  0  0  0  0
 ·  ·  ·  0  0  0  0  0  0  O  ·  ·  ·
 ·  ·  ·  0  0  0  0  0  0  0  O  ·  ·
 ·  ·  ·  0  0  0  0  0  0  0  0  O  ·
 ·  ·  ·  0  0  0  0  0  0  0  0  0  O
```

x_m is the square table with m rows and m columns such that:
- the main diagonal contains only 0's,
- above the main diagonal, there is a sequence of $\lfloor \log m \rfloor$ standing side by side squares such that: the bottom left corner of the i-th square stands just above the main diagonal in the cell $x_m(2^{i-1}, 2^{i-1}+1)$ and the upper right corner in $x_m(1, 2^i)$, the edge of each square consists of 0's and the interior of 1's. If m is not a power of 2, then after the last square there is a rectangle with the bottom left corner in $x_m(2^{\lfloor \log m \rfloor}, 2^{\lfloor \log m \rfloor}+1)$ and upper right in $x_m(1, m)$. In all other cells above the diagonal stands 1.
- below the main diagonal for each i, $1 \leq i \leq \lfloor \log m \rfloor$, there are 1's in the i-th column from $x_m(2^{i-1}+1, i)$ to $x_m(m, i)$. There is 0 in all other cells below the main diagonal. Note that the last 0 in the i-th column stands in the same row as the bottom left corner of the i-th square above the main diagonal.

Now we shall show how the finite automaton A accepts LOG. First it checks whether on the main diagonal of an input x there are only 0's and that above the diagonal there is a sequence of rectangles with 0's on the edge and 1's inside. The first rectangle should be a square consisting of one cell $x(1,2)$. The rectangles should be adjacent to each other and to the upper edge of x and the bottom left corner of each rectangle should stand just above the main diagonal of x. Next it checks if all rectangles, excepting possibly the last one, are squares, by going through the main diagonal of each rectangle. Note that if all the above conditions are satisfied, then the side of each square is twice as long as the side of the previous square and the bottom left corner of the i-th square stands in $x(2^{i-1}, 2^{i-1}+1)$. Then A checks if below the main diagonal on the bottom of the first few columns there are 1's and if the number of 1's in each column is less than in the previous one standing to the left. Next A checks if the number of columns with 1's below the main diagonal is equal to the number of squares above and if the last 0 in the i-th column is in the same row as the bottom left corner of the i-th square. To do this, A checks one by one if the last 0 in the i-th column for $i \geq 3$ (the first two columns can be checked separately), stands in the same row as the bottom left corner of some square, and if the bottom left corner of each square stands in the same row as the last 0 in some column. Note that if $i \geq 3$, then the bottom left corner of the square can easily be recognized.   □

Now we are ready to prove that the functions $\log^* n$ and $\log^{(k)} n$ for every k, are space constructible by two-dimensional Turing machines.

Theorem 15.1.5.   The functions $\log^* n$ and $\log^{(k)} n$ for every k, are 2-space constructible.

Proof.   First we show that for every k, there exists a two-dimensional deterministic Turing machine M that constructs the function $\log^{(k)} n$. The input alphabet of M consists of three symbols $\{0, 1, 2\}$. We should show that for every n, there is a square input, say $G(n)$, with n rows and n columns on which M uses exactly $\log^{(k)} n$ space. To construct $G(n)$, we use the language LOG described in Lemma 15.1.4. The $\lfloor \frac{n}{2} \rfloor + 1$ row and the $\lfloor \frac{n}{2} \rfloor + 1$ column of $G(n)$ contains only 2's. This row and column divide $G(n)$ into four rectangles. The upper left rectangle belongs to LOG, in its last row there are 1's in the first $\lfloor \log n \rfloor - 1$ cells. Just below these 1's in the bottom left rectangle we build the second square with side-length $\lfloor \log n \rfloor$ that again belongs to LOG. This square is surrounded by 2's. Below the 1's in the last row of the second square we build the third square from LOG with side-length $\log \log n$, and so on. There should be k squares from LOG one below the other. The last square contains $\log^{(k)} n$ 1's in its last row. Since the language LOG is accepted by the finite automaton A, M can check without using any space if the input is equal to $G(n)$ for some n. If this is the case, M marks on its work tape as many cells as there are 1's in the last row of the k-th square.

The function $\log^* n$ can be constructed in a similar way, but then in $G(n)$ the squares from LOG are built until the last of them contains in its last row only one 1, and M counts the number of squares.   □

Up to now, we have been considering functions constructed by Turing machines with square inputs. It is easy to see that if inputs of Turing machines are not restricted to squares, then we can construct the same functions as the functions of the minimum of the two side-lengths of the input. More precisely, there exists a two-dimensional deterministic Turing machine which uses exactly $\log \log(\min(m, n))$ space on every input with m rows and n columns. Similarly, there exists a two-dimensional deterministic Turing machine which for every m and n, uses $\log^*(\min(m, n))$ (or $\log^{(k)}(\min(m, n))$) space on some input with m rows and n columns.

Note that it is not possible to use maximum instead of minimum because otherwise one could construct these functions in one dimension by using inputs with one row.

15.2. Inkdot Turing Machines

In this section we consider the 1-inkdot Turing machines and present a result of Hartmanis and Ranjan (1989) that the function $\log \log n$ can be fully space constructed by a nondeterministic 1-inkdot Turing machine, and that for 1-inkdot Turing machines, nondeterministic and deterministic sublogarithmic space complexity classes are not equal.

Definition 15.2.1. *A 1-inkdot Turing machine is a standard Turing machine with the additional power of marking 1 cell on the input tape (with an inkdot). This cell is marked once and for all (no erasing) and no more than one dot of ink is available. The action of the machine depends on the current state, the currently scanned input and work tape symbols, and the presence of the inkdot on the currently scanned cell. The action consists of moving the heads, making changes on the work tape, and changing the state of the finite control; in addition, the inkdot may be used to mark currently scanned cell on the input tape if it has not already been used.*

Definition 15.2.2.
- $DSPACE^{dot}[L(n)] = \{L \mid L$ is accepted by an $L(n)$ *space-bounded deterministic 1-inkdot Turing machine* $\}$.
- $NSPACE^{dot}[L(n)] = \{L \mid L$ is accepted by an $L(n)$ *space-bounded nondeterministic 1-inkdot Turing machine*$\}$.

It is obvious that if $L(n) \geq \log n$, then the inkdot does not increase the power of neither deterministic nor nondeterministic Turing machines. Thus we have.

Theorem 15.2.3. If $L(n) \geq \log n$, then

$$DSPACE^{dot}[L(n)] = DSPACE[L(n)] \qquad \text{and}$$
$$NSPACE^{dot}[L(n)] = NSPACE[L(n)].$$

But also for arbitrary function $L(n)$, the inkdot does not increase the power of strongly $L(n)$ space-bounded deterministic Turing machines (Hartmanis and Ranjan 1989).

Theorem 15.2.4. For an arbitrary function $L(n)$,

$$strong\text{-}DSPACE^{dot}[L(n)] = strong\text{-}DSPACE[L(n)].$$

Proof. Let M_{dot} be a strongly $L(n)$ space-bounded 1-inkdot machine. We shall show that there is a normal strongly $L(n)$ space-bounded Turing machine M which simulates M_{dot} without using the inkdot. The simulation of M_{dot} by M is straight forward until (if at all) the dot is placed on the input tape by M_{dot}. At this time, M records β –the internal configuration of M_{dot}, and σ –the symbol being scanned on the input tape. Note that M cannot remember where the dot was placed on the input tape if it is forced to move away from the dot. In the following simulation, the dot's position will be recomputed again and again.

After the placement of the dot by M_{dot}, the forward simulation by M continues until the symbol σ is scanned (it could have a dot mark). In this case, β' –the current internal configuration of M_{dot} is recorded and a depth-first, backwards search is initiated to determine if we can back M_{dot} from the internal configuration β to the starting configuration. The search is performed similarly as it was described in the proof of Theorem 7.2.1. If the starting configuration has been reached, then M simulates M_{dot} forward until M_{dot} reaches the dot operation; now we are back on the cell with the dot on it; Then M switches to the stored internal configuration β' of M_{dot} and continues the forward simulation of M_{dot}. If the starting configuration cannot be reached, then we know that the dot was not placed on this cell and in this case the depth first search will halt with head in the position from where it started. Now M continues the simulation of M_{dot} from the configuration β', knowing that there is no inkdot on this cell.   □

Now we present the theorem of (Hartmanis and Ranjan 1989) that a nondeterministic 1-inkdot Turing machine can fully construct the function $\log \log n$.

Theorem 15.2.5. The function $\log \log n$ is fully space constructible by a nondeterministic 1-inkdot Turing machine. That is, there is a strongly $O(\log \log n)$ space-bounded nondeterministic 1-inkdot machine M such that on every input of length n, it reaches a configuration which is exactly $\log \log n$ space-bounded.

Proof. On an input of length n, M places the inkdot somewhere in the input, say at the cell k. Then M computes the binary description of

$$F(k) = \min\{i \mid i \text{ does not divide } k\}.$$

By Lemma 4.1.2(d), there is a constant $d > 0$ such that $F(k) \leq d \log k$. Thus, M is strongly $O(\log \log n)$ space-bounded.

Consider now

$$k = lcm\{i \mid i \leq c \log n\},$$

where $lcmA$ denotes the least common multiple of elements of A and c is a constant $c > 0$. By Lemma 4.1.2(b),

$$k \leq c_2^{c \log n}$$

and there is $c > 0$ such that $k \leq n$. On the other hand, by Lemma 4.1.2(f)

$$F(k) > c \log n.$$

Hence, there is a computation of M that uses $\log F(n) \geq \log(c \log n)$ space. Thus, there is a constant $d_1 > 0$ such that M uses at least $d_1 \log \log n$ space on every input of length n. At the end, if necessary (if $d_1 < 1$), M multiplies the number of cells on the work tape by a constant to get $\log \log n$.   □

The fact that $\log \log n$ is fully space constructible by a nondeterministic 1-inkdot Turing machine and not by deterministic one, implies that for 1-inkdot Turing machines, nondeterministic and deterministic space complexity classes are not equal. This is formalized in the following theorem (Hartmanis and Ranjan 1989).

Theorem 15.2.6.   There is a language A such that
- $A \in$ *strong-NSPACE* $^{dot}[\log \log n]$ and
- $A \notin$ *strong-DSPACE* $^{dot}[L(n)]$ for any $L(n) = o(\log n)$.

As an immediate corollary we have the following theorem.

Theorem 15.2.7.   If $L(n) \geq \log \log n$ and $L(n) = o(\log n)$, then

$$strong\text{-}DSPACE^{dot}[L(n)] \subsetneqq strong\text{-}NSPACE^{dot}[L(n)].$$

Proof of Theorem 15.2.6. Consider the language

$$A = \{a^k b^m \mid k \neq m\}.$$

By Lemma 8.3.3,

$$A \notin strong\text{-}DSPACE[L(n)]$$

and by Theorem 15.2.4,

$$A \notin strong\text{-}DSPACE^{dot}[L(n)].$$

Now we shall show that there is a strongly $\log \log n$ space-bounded nondeterministic 1-inkdot Turing machine M accepting A. On any input of the form $w = a^k b^m$, M first marks off $\log \log(k+m)$ cells on the work tape. This is possible, since by Theorem 15.2.5, $\log \log n$ is fully space constructible by a nondeterministic 1-inkdot machine. Next, using marked space M checks if $k \neq m \pmod{r}$ for some $r \leq d \log(k+m)$, and accepts if this is the case. Remember that by Lemma 4.1.2, there is a constant d such that for every k and m, $k \neq m$, there is $r \leq d \log(k+m)$ such that $k \neq m \pmod{r}$.   □

15.3. 1-pebble Turing Machines

In this section we consider Turing machines equipped with one pebble which can be placed on and removed from the input tape. The 1-pebble finite automata accept only regular languages. On the other hand, the pebble is redundant when the space used is $\log n$ or more because then the position of the pebble can be remembered in binary (cf. Section 3.2).

Chang et. al. (1986) proved that $\log \log n$ is fully space constructed by a deterministic 1-pebble Turing machine. This means that the pebble increases the power of deterministic Turing machines with sublogarithmic space. They also showed that $\log \log n$ is the lower bound for accepting non-regular languages by 1-pebble (and 1-inkdot) Turing machines.

Definition 15.3.1. *A 1-pebble Turing machine is a Turing machine with a two-way read-only input tape, a two-way read-write work tape, and a pebble which can be placed on and removed from the input tape. The action of the Turing machine depends on the current state, the currently scanned input and work tape symbols, and the presence of the pebble on the currently scanned cell. The action consists of moving the heads, making changes on the work tape, changing the state of the finite control, and picking up or placing (or none) the pebble on the currently scanned cell on the input tape.*

Definition 15.3.2.
- $DSPACE^{peb}[L(n)] = \{L \mid L$ is accepted by an $L(n)$ space-bounded deterministic 1-pebble Turing machine$\}$.
- $NSPACE^{peb}[L(n)] = \{L \mid L$ is accepted by an $L(n)$ space-bounded non-deterministic 1-pebble Turing machine$\}$.

First we present the theorem of (Chang et. al. 1986) that $\log \log n$ is fully space constructible by a deterministic 1-pebble Turing machine.

Theorem 15.3.3. The function $\log \log n$ is fully space constructible by a deterministic 1-pebble Turing machine. That is, there is deterministic 1-pebble machine M that uses exactly $\log \log n$ space on every input of length n.

Proof. On an input of length n, M places the pebble at a position k and finds the binary description of

$$F(k) = \min\{i \mid i \text{ does not divide } k\}.$$

It repeats this procedure for $k = 1, 2, \ldots, n$. In much the same way as in the proof of Theorem 15.2.5 one can show that M fully constructs $\log \log n$. $\square$

As a corollary we have the following theorem.

Theorem 15.3.4. If $L(n) \geq \log \log n$ and $L(n) = o(\log n)$, then

$$DSPACE[L(n)] \subsetneq DSPACE^{peb}[L(n)].$$

Proof. Consider the language

$$A = \{a^k b^k \mid k \geq 0\}.$$

By Lemma 6.2.5,

$$A \notin DSPACE[L(n)].$$

On the other hand $\log\log n$ is fully space constructible by a deterministic 1-pebble Turing machine and just as in the proof of Theorem 15.2.5 one can show that

$$A \in DSPACE^{peb}[\log\log n].$$

□

Now we sketch the proof that $\log\log n$ is the lower bound for accepting non-regular languages by 1-pebble Turing machines, and so by 1-inkdot machines (Chang et. al 1986).

Theorem 15.3.5. Let M be an $L(n)$ space-bounded 1-pebble deterministic or nondeterministic Turing machine and $L(n) = o(\log\log n)$, then M accepts only a regular language.

Proof. The proof is similar to that of Theorem 5.1.3. We need only to modify the equivalence relation defined in Definition 5.1.1. Let M be an weakly $L(n)$ space-bounded deterministic or nondeterministic 1-pebble Turing machine with an input alphabet Σ, and let k be a natural number. Using counting argument similar as in the proof of Theorem 5.1.3 one can show that if $L(n) = o(\log\log n)$, then there are x, y, $z \in \Sigma^*$ such that $yz \equiv_k z$ and $x \simeq_k xy$, where $\equiv_k$ is the equivalence relation defined in Definition 5.1.1 and the relation $\simeq_k$ is defined analogously as $\equiv_k$ but for configuration entering the words from the right instead from the left. Moreover, for every internal configurations β_1, β_2,

- there is a k space-bounded computation of M on xyz that starting in β_1 with the pebble at the first letter of z reaches β_2 with the pebble just after it leaves z

if and only if

- there is a k space-bounded computation of M on xyz that starting in β_1 with the pebble at the first letter of yz reaches β_2 with the pebble just after it leaves yz.

During the computation from β_1 to β_2, M can leave z or yz but without the pebble.

As in Lemma 5.1.2 one can prove that in this case M accepts xyz in k space if and only if M accepts xz in k space.   □

Corollary 15.3.6. The 1-pebble finite automata accept only regular languages.

15.4. Demon Turing Machines

To emphasize the importance of the constructibility of tape bounds Hartmanis and Ranjan (1989) proposed to consider a demon Turing machine.

Definition 15.4.1. *A demon $L(n)$ Turing machine is a Turing machine that has $L(n)$ cells marked on its work tape for every input of length n, and cannot use more space than $L(n)$.*

Definition 15.4.2.
- $DSPACE^{dem}[L(n)] = \{L \mid L$ is accepted by an $L(n)$ space-bounded deterministic demon Turing machine$\}$.
- $NSPACE^{dem}[L(n)] = \{L \mid L$ is accepted by an $L(n)$ space-bounded nondeterministic demon Turing machine$\}$.

The demon Turing machine has $L(n)$ space marked automatically on any input of length n, and it does not need any space constructibility. It is easy to see that a demon $L(n)$ Turing machine is equivalent to a standard $L(n)$ space-bounded Turing machine if the function $L(n)$ is fully space constructible. Hence, they are equivalent for all "normal" functions above logarithm.

Consider two languages:
$$\{a^k b^k \mid k \geq 0\}$$
and
$$\{w\#w \mid w \in \{a, b\}^*\}.$$

They both require $O(\log n)$ space to be accepted by a standard Turing machine.

Below we present results of Hartmanis and Ranjan (1989) who showed that $\{a^k b^k \mid k \geq 0\}$ is accessible in $O(\log \log n)$ space by a demon Turing machine, but $\{w\#w \mid w \in \{a, b\}^*\}$ requires $O(\log n)$ space. This very clearly shows that the lower bound, $O(\log n)$, on the amount of space required to recognize $\{a^k b^k \mid k \geq 0\}$ is purely due to space constructibility reasons, whereas $\{w\#w \mid w \in \{a, b\}^*\}$ requires $O(\log n)$ space independently of any constructibility properties of the Turing machine.

Theorem 15.4.3.
(a) $\{a^k b^k \mid k \geq 0\} \in DSPACE^{dem}[\log \log n]$ and
(b) $\{a^k b^k \mid k \geq 0\} \notin NSPACE^{dem}[L(n)]$ if $L(n) = o(\log \log n)$.

Proof.
(a) Similarly to the proof of Theorem 15.2.6 using the fact that $\log \log |w|$ space is now marked off automatically.

(b) Suppose that the language $\{a^k b^k \mid k \geq 0\}$ can be accepted by some $L(n)$ space-bounded nondeterministic demon Turing machine M. Let w be some input of the form $w = a^k b^k$ and let ℓ be the space used by M on w, i.e. $\ell = L(2k)$. Consider the equivalence relation $\equiv_\ell$ defined in Definition 5.1.1. It is easy to see that if $L(n) = o(\log \log n)$, then for k large enough, k is greater than the number of equivalence classes of the relation $\equiv_\ell$. Thus, there exist two numbers i and j, such that $1 \leq i < j \leq k$ and $b^i \equiv_\ell b^j$. In that case, by Lemma 5.1.2, also the word $a^k b^{k+i-j}$ is accepted by M, a contradiction.   □

Theorem 15.4.4. If $L(n) = o(\log n)$, then

$$\{w \# w \mid w \in \{a, b\}^*\} \notin NSPACE^{dem}[L(n)].$$

Proof. The theorem is actually proved in the proof of Theorem 11.2.1. One should only consider $L(n) = \log n$.   □

References

Alberts, M. (1985): *Space complexity of alternating Turing machines*, in: L. Budach, ed., Fundamentals of Computation Theory, LNCS 199 (Springer, Berlin) 1–7.

Alt, H. and Melhorn, K. (1975): *A language over a one symbol alphabet requiring only $O(\log \log n)$ space*, SIGACT Newslett. (Nov./Dec. 1975) 31–33.

Balcázar, J. L., Díaz, J., and Gabarró, J. (1988): *Structural Complexity I*, EATCS Monographs on Theoretical Computer Science, Vol. 11, Springer, Berlin.

Birget, J.-C. (1989): *Proof of a conjecture of R. Kannan, and generalizations*, Univ. of Nebraska - Lincoln, Dept of Computer Sci. and Engineering, Report Series #77, December 1988 (revised, May 1989).

Braunmühr, B. von, Gengler, R., and Rettinger, R. (1994): *The alternating hierarchy for machines with sublogarithmic space is infinite*, in: P. Enjalbert, E.W. Mayr, K. W. Wagner eds., Proc. of STACS'94, LNCS 775 (Springer, Berlin) 85–96.

Chandra, A. K., Kozen D. C., and Stockmeyer, L. J. (1981): *Alternation*, J.ACM 28, 114–133.

Chang, J. H., Ibarra, O. H., Palis, M. A., and Ravikumar, B. (1986): *On pebble automata*, TCS 44, 111–121.

Chang, J. H., Ibarra, O. H., Ravikumar, B., and Berman, L. (1987): *Some observations concerning alternating Turing machines using small space*, IPL 25 1–9 [Erratum IPL 27 (1988) p.53].

Freedman, A. R. and Ladner, R. E. (1975): *Space bounds for processing contentless inputs*, J.CSS 11, 118–128.

Freivalds, R. (1979): *On the worktime of deterministic and nondeterministic Turing machines* (in Russian), Latvijskij Matematiceskij Eshegodnik 23, 158–165.

Geffert, V. (1991): *Nondeterministic computations in sublogarithmic space and space constructibility*, SIAM J. Comput. 20, 484–498.

Hartmanis, J. (1972): *On non-determinancy in simple computing devices*, Acta Informatica 1, 336–344.

Hartmanis, J. and Berman, L. (1976): *On tape bounds for single-letter alphabet language processing*, TCS 3, 213–224.

Hartmanis, J. and Ranjan, D. (1989): *Space bounded computations: review and new separation results*, in: A. Kreczmar, G. Mirkowska eds., Proc. of Mathematical Foundations of Computer Science, LNCS 379 (Springer, Berlin) 49–66.

Hopcroft, J. E. and Ullman, J. D. (1967): *Some results on tape bounded Turing machines*, J.ACM 16, 1, 168–177.

Hopcroft, J. E. and Ullman, J. D. (1979): *Introduction to Automata Theory, Languages, and Computation*, Addison-Wesley.

Ibarra, O. H. and Ravikumar, B. (1988): *Sublogarithmic space Turing machines, nonuniform space complexity, and closure properties*, Math. Syst. Th. 21, 1–17.

Immerman, N. (1988): *Nondeterministic space is closed under complement*, SIAM J. Comput. 17, 935–938.

Ito, A., Inoue, K., and Takanami, I. (1987): *A note on alternating Turing machines using small space*, The Transactions of the IEICE, Vol. E 70, No. 10, 990–996.

Iwama, K. (1993): $ASPACE(o(\log\log n))$ is regular, SIAM J. Comput. 22, 136–146.

Kannan, R. (1983):Alternation and the power of nondeterminism, 15-th STOC, 344–346.

Lewis II, P. M., Stearns, R. E., and Hartmanis, J. (1965): Memory bounds for recognition of context-free and context-sensitive languages, in: IEEE Conf. Rec. on Switching Circuit Theory and Logical Design, NY, 191–202.

Liśkiewicz, M. and Reischuk, R. (1994): The complexity world below logarithmic space, Proc. of Structure in Complexity Theory, Amsterdam.

Michel, P. (1992): A survey of space complexity, TCS 101, 99–132.

Monien, B. (1981): On the LBA problem, in: F. Gécseg, ed., Fundamentals of Computation Theory, LNCS 117 (Springer, Berlin) 265–280.

Monien, B. and Sudborough, I. H. (1982): On eliminating nondeterminism from Turing machines which use less than logarithm worktape space, TCS 21, 237–253.

Narkiewicz, W. (1983): Number Theory (translated by S. Kanemitsu), World Scientific Publishing Co.

Savitch, W. J. (1970): Relationships between nondeterministic and deterministic tape complexities, J.CSS, 4, 177–192.

Seiferas, J. I. (1977): Techniques for separating space complexity classes, J.CSS 14, 73–99.

Sipser, M. (1980): Halting space-bounded computations, TCS 10, 335–338.

Stearns, R. E., Hartmanis, J., and Lewis II, P. M. (1965): Hierarchies of memory limited computations, in: IEEE Conf. Rec. on Switching Circuit Theory and Logical Design, NY, 179–190.

Szelepcsényi, R. (1988): The method of forced enumeration for nondeterministic automata, Acta Informatica 26, 279–284.

Szepietowski, A. (1987): There are no fully space constructible functions between $\log\log n$ and $\log n$, IPL 24, 361–362.

Szepietowski, A. (1988): Remarks on languages acceptable in $\log\log n$ space, IPL 27, 201–203.

Szepietowski, A. (1988a) Alternating space below $\log n$ is more powerful than nondeterministic one, Schriften zur Informatik und Angewandten Mathematik, Bericht Nr. 137, September 1988, Rheinisch-Westfälische Technische Hochschule Aachen, Templergraben 55, 5100 Aachen.

Szepietowski, A. (1989): Some remarks on the alternating hierarchy and closure under complement for sublogarithmic space, IPL 33, 73–78.

Szepietowski, A. (1989a): Some notes on strong and weak $\log\log n$ space complexity, IPL 33, 109–112.

Szepietowski, A. (1990): If deterministic and nondeterministic space complexities are equal for $\log\log n$ then they are also equal for $\log n$, TCS 74, 115–119.

Szepietowski, A. (1990a): Weak mode of space complexity can be used in the proof that $\left[DSPACE(\log\log n) = NSPACE(\log\log n)\right] \Rightarrow \left[L = NL\right]$, Bulletin of the European Association for Theoretical Computer Science, EATCS, No. 40, February 1990, 266–269.

Szepietowski, A. (1992): On space functions constructed by two-dimensional Turing machines, Information Sciences 60, 177–183.

Subject Index

Accepting computation, 10
Accepting configuration, 10
Accepting state, 8
Accepting subtree, 30
Accepting tree, 12
Accessible configuration, 10
Alphabet, 7
Alternating Turing machine, 12
Alternation-bounded Turing
    machine, 12

Blank symbol, 8
Bounded language, 7

Cell, 7
Complement, 7
Computation, 10
Computation tree, 12
Configuration, 9
Concatenation, 7
Counter, 17

Demon Turing machine, 109
Deterministic Turing machine, 9
Difference, 7

Empty word, 7
Endmarker, 8
Existential state, 12
Existential configuration, 12

Final configuration, 10
Fully space constructible
    function, 37
Fully 2-space constructible
    function, 100

Halting Turing machine, 48

Immediate successor, 9
Inclusion, 7
Independent complement, 95
Independent simulation, 95
Initial configuration, 9
Initial state, 8
Input, 8
Input head, 8
Input tape, 8
Input tape alphabet, 8
Internal configuration, 11
Intersection, 7

Kleene closure, 7
k-equivalent, 28
k-pebble automaton, 16

Language, 7
Language accepted by Turing
    machine, 10
Left endmarker, 8
Length, 7
Letter, 7
log-space reduction, 19
Middle space-bounded Turing
    machine, 13

Nondeterministic Turing machine, 7,
Nondeterministically fully space
    constructible function, 37
$NSPACE(\log n)$-complete language, 19

One-way Turing machine, 9
1-inkdot Turing machine, 104
1-pebble Turing machine, 107

Padding, 68, 70
Power set, 7

Proper inclusion, 7
Pumping lemma, 22
Push-down tape, 16
Π_k-alternating Turing machine, 13
Recognizing Turing machine, 10
Right endmarker, 8

Set of states, 7
Space, 13
Space-bounded Turing machine, 13
Space complexity class, 14
Space constructible function, 37
State, 7
Step, 8
Strongly space-bounded Turing
    machine, 13
Symbol, 7
Σ_k-alternating Turing machine, 13

Tally language, 7
Transition function, 8
Tree of accessible configurations, 10
Transducer 19
Turing machine, 7
Two-dimensional tape, 100
Two-dimensional Turing machine, 100
2-space constructible function, 100

Union, 7
Universal configuration, 12
Universal state, 12

Weakly space-bounded Turing
    machine, 13
Word, 7
Work head, 7
Work tape, 7
Work tape alphabet, 8

Symbol Index

λ, 7
$|w|$, 7
$w(i)$, 7
Σ^*, 7
$\$$, 8
$\not\$$, 8
$A \cup B$, 7
$A \cap B$, 7
$A - B$, 7
$A^c = \Sigma^* - A$, 7
AB, 7
A^n, 7
A^*, 7
$a^* = \{a\}^*$, 7
$A \subseteq B$, 7
$A \subsetneqq B$, 7
$\mathcal{P}(A)$, 7
co-$\mathcal{C}$, 7
$lcmA$, 22

$\Sigma^{(2)}$, 100
$x(i, j)$, 100

$O(g(n))$, 14
$o(g(n))$, 14
$f(n) \ll g(n)$, 14
$\log n$, 14
$\log^{(k)} n$, 14
$\log^* n$, 14

$Pad(G, M)$, 68
$P(G)$, 70

$\equiv_k$, 28
$\cong_k$, 84
$\simeq_k$, 108

$DSPACE[L(n)]$, 14
$NSPACE[L(n)]$, 14
$ASPACE[L(n)]$, 14
$\Pi_k\text{-}SPACE[L(n)]$, 14
$\Sigma_k\text{-}SPACE[L(n)]$, 14

strong, 14
weak, 14
middle, 14
one-way, 14

$SEC_{x|y}(T)$, 30
$CSR(T)$, 30
$SP(T)$, 30
$S^k_{x|y}(r)$, 30
$S^k_{x|y}$, 31
$MINS_{x|y}(r)$, 31
$MINS_{x|y}$, 32

$DSPACE^{dot}[L(n)]$, 104
$NSPACE^{dot}[L(n)]$, 104
$DSPACE^{peb}[L(n)]$, 107
$NSPACE^{peb}[L(n)]$, 107
$DSPACE^{dem}[L(n)]$, 109
$NSPACE^{dem}[L(n)]$, 109

Springer-Verlag
and the Environment

We at Springer-Verlag firmly believe that an international science publisher has a special obligation to the environment, and our corporate policies consistently reflect this conviction.

We also expect our business partners – paper mills, printers, packaging manufacturers, etc. – to commit themselves to using environmentally friendly materials and production processes.

The paper in this book is made from low- or no-chlorine pulp and is acid free, in conformance with international standards for paper permanency.

Printing: Weihert-Druck GmbH, Darmstadt
Binding: Theo Gansert Buchbinderei GmbH, Weinheim

Lecture Notes in Computer Science

For information about Vols. 1–762
please contact your bookseller or Springer-Verlag

Vol. 763: F. Pichler, R. Moreno Díaz (Eds.), Computer Aided Systems Theory – EUROCAST '93. Proceedings, 1993. IX, 451 pages. 1994.

Vol. 764: G. Wagner, Vivid Logic. XII, 148 pages. 1994. (Subseries LNAI).

Vol. 765: T. Helleseth (Ed.), Advances in Cryptology – EUROCRYPT '93. Proceedings, 1993. X, 467 pages. 1994.

Vol. 766: P. R. Van Loocke, The Dynamics of Concepts. XI, 340 pages. 1994. (Subseries LNAI).

Vol. 767: M. Gogolla, An Extended Entity-Relationship Model. X, 136 pages. 1994.

Vol. 768: U. Banerjee, D. Gelernter, A. Nicolau, D. Padua (Eds.), Languages and Compilers for Parallel Computing. Proceedings, 1993. XI, 655 pages. 1994.

Vol. 769: J. L. Nazareth, The Newton-Cauchy Framework. XII, 101 pages. 1994.

Vol. 770: P. Haddawy (Representing Plans Under Uncertainty. X, 129 pages. 1994. (Subseries LNAI).

Vol. 771: G. Tomas, C. W. Ueberhuber, Visualization of Scientific Parallel Programs. XI, 310 pages. 1994.

Vol. 772: B. C. Warboys (Ed.),Software Process Technology. Proceedings, 1994. IX, 275 pages. 1994.

Vol. 773: D. R. Stinson (Ed.), Advances in Cryptology – CRYPTO '93. Proceedings, 1993. X, 492 pages. 1994.

Vol. 774: M. Banâtre, P. A. Lee (Eds.), Hardware and Software Architectures for Fault Tolerance. XIII, 311 pages. 1994.

Vol. 775: P. Enjalbert, E. W. Mayr, K. W. Wagner (Eds.), STACS 94. Proceedings, 1994. XIV, 782 pages. 1994.

Vol. 776: H. J. Schneider, H. Ehrig (Eds.), Graph Transformations in Computer Science. Proceedings, 1993. VIII, 395 pages. 1994.

Vol. 777: K. von Luck, H. Marburger (Eds.), Management and Processing of Complex Data Structures. Proceedings, 1994. VII, 220 pages. 1994.

Vol. 778: M. Bonuccelli, P. Crescenzi, R. Petreschi (Eds.), Algorithms and Complexity. Proceedings, 1994. VIII, 222 pages. 1994.

Vol. 779: M. Jarke, J. Bubenko, K. Jeffery (Eds.), Advances in Database Technology — EDBT '94. Proceedings, 1994. XII, 406 pages. 1994.

Vol. 780: J. J. Joyce, C.-J. H. Seger (Eds.), Higher Order Logic Theorem Proving and Its Applications. Proceedings, 1993. X, 518 pages. 1994.

Vol. 781: G. Cohen, S. Litsyn, A. Lobstein, G. Zémor (Eds.), Algebraic Coding. Proceedings, 1993. XII, 326 pages. 1994.

Vol. 782: J. Gutknecht (Ed.), Programming Languages and System Architectures. Proceedings, 1994. X, 344 pages. 1994.

Vol. 783: C. G. Günther (Ed.), Mobile Communications. Proceedings, 1994. XVI, 564 pages. 1994.

Vol. 784: F. Bergadano, L. De Raedt (Eds.), Machine Learning: ECML-94. Proceedings, 1994. XI, 439 pages. 1994. (Subseries LNAI).

Vol. 785: H. Ehrig, F. Orejas (Eds.), Recent Trends in Data Type Specification. Proceedings, 1992. VIII, 350 pages. 1994.

Vol. 786: P. A. Fritzson (Ed.), Compiler Construction. Proceedings, 1994. XI, 451 pages. 1994.

Vol. 787: S. Tison (Ed.), Trees in Algebra and Programming – CAAP '94. Proceedings, 1994. X, 351 pages. 1994.

Vol. 788: D. Sannella (Ed.), Programming Languages and Systems – ESOP '94. Proceedings, 1994. VIII, 516 pages. 1994.

Vol. 789: M. Hagiya, J. C. Mitchell (Eds.), Theoretical Aspects of Computer Software. Proceedings, 1994. XI, 887 pages. 1994.

Vol. 790: J. van Leeuwen (Ed.), Graph-Theoretic Concepts in Computer Science. Proceedings, 1993. IX, 431 pages. 1994.

Vol. 791: R. Guerraoui, O. Nierstrasz, M. Riveill (Eds.), Object-Based Distributed Programming. Proceedings, 1993. VII, 262 pages. 1994.

Vol. 792: N. D. Jones, M. Hagiya, M. Sato (Eds.), Logic, Language and Computation. XII, 269 pages. 1994.

Vol. 793: T. A. Gulliver, N. P. Secord (Eds.), Information Theory and Applications. Proceedings, 1993. XI, 394 pages. 1994.

Vol. 794: G. Haring, G. Kotsis (Eds.), Computer Performance Evaluation. Proceedings, 1994. X, 464 pages. 1994.

Vol. 795: W. A. Hunt, Jr., FM8501: A Verified Microprocessor. XIII, 333 pages. 1994.

Vol. 796: W. Gentzsch, U. Harms (Eds.), High-Performance Computing and Networking. Proceedings, 1994, Vol. I. XXI, 453 pages. 1994.

Vol. 797: W. Gentzsch, U. Harms (Eds.), High-Performance Computing and Networking. Proceedings, 1994, Vol. II. XXII, 519 pages. 1994.

Vol. 798: R. Dyckhoff (Ed.), Extensions of Logic Programming. Proceedings, 1993. VIII, 362 pages. 1994.

Vol. 799: M. P. Singh, Multiagent Systems. XXIII, 168 pages. 1994. (Subseries LNAI).

Vol. 800: J.-O. Eklundh (Ed.), Computer Vision – ECCV '94. Proceedings 1994, Vol. I. XVIII, 603 pages. 1994.

Vol. 801: J.-O. Eklundh (Ed.), Computer Vision – ECCV '94. Proceedings 1994, Vol. II. XV, 485 pages. 1994.

Vol. 802: S. Brookes, M. Main, A. Melton, M. Mislove, D. Schmidt (Eds.), Mathematical Foundations of Programming Semantics. Proceedings, 1993. IX, 647 pages. 1994.

Vol. 803: J. W. de Bakker, W.-P. de Roever, G. Rozenberg (Eds.), A Decade of Concurrency. Proceedings, 1993. VII, 683 pages. 1994.

Vol. 804: D. Hernández, Qualitative Representation of Spatial Knowledge. IX, 202 pages. 1994. (Subseries LNAI).

Vol. 805: M. Cosnard, A. Ferreira, J. Peters (Eds.), Parallel and Distributed Computing. Proceedings, 1994. X, 280 pages. 1994.

Vol. 806: H. Barendregt, T. Nipkow (Eds.), Types for Proofs and Programs. VIII, 383 pages. 1994.

Vol. 807: M. Crochemore, D. Gusfield (Eds.), Combinatorial Pattern Matching. Proceedings, 1994. VIII, 326 pages. 1994.

Vol. 808: M. Masuch, L. Pólos (Eds.), Knowledge Representation and Reasoning Under Uncertainty. VII, 237 pages. 1994. (Subseries LNAI).

Vol. 809: R. Anderson (Ed.), Fast Software Encryption. Proceedings, 1993. IX, 223 pages. 1994.

Vol. 810: G. Lakemeyer, B. Nebel (Eds.), Foundations of Knowledge Representation and Reasoning. VIII, 355 pages. 1994. (Subseries LNAI).

Vol. 811: G. Wijers, S. Brinkkemper, T. Wasserman (Eds.), Advanced Information Systems Engineering. Proceedings, 1994. XI, 420 pages. 1994.

Vol. 812: J. Karhumäki, H. Maurer, G. Rozenberg (Eds.), Results and Trends in Theoretical Computer Science. Proceedings, 1994. X, 445 pages. 1994.

Vol. 813: A. Nerode, Yu. N. Matiyasevich (Eds.), Logical Foundations of Computer Science. Proceedings, 1994. IX, 392 pages. 1994.

Vol. 814: A. Bundy (Ed.), Automated Deduction—CADE-12. Proceedings, 1994. XVI, 848 pages. 1994. (Subseries LNAI).

Vol. 815: R. Valette (Ed.), Application and Theory of Petri Nets 1994. Proceedings. IX, 587 pages. 1994.

Vol. 816: J. Heering, K. Meinke, B. Möller, T. Nipkow (Eds.), Higher-Order Algebra, Logic, and Term Rewriting. Proceedings, 1993. VII, 344 pages. 1994.

Vol. 817: C. Halatsis, D. Maritsas, G. Philokyprou, S. Theodoridis (Eds.), PARLE '94. Parallel Architectures and Languages Europe. Proceedings, 1994. XV, 837 pages. 1994.

Vol. 818: D. L. Dill (Ed.), Computer Aided Verification. Proceedings, 1994. IX, 480 pages. 1994.

Vol. 819: W. Litwin, T. Risch (Eds.), Applications of Databases. Proceedings, 1994. XII, 471 pages. 1994.

Vol. 820: S. Abiteboul, E. Shamir (Eds.), Automata, Languages and Programming. Proceedings, 1994. XIII, 644 pages. 1994.

Vol. 821: M. Tokoro, R. Pareschi (Eds.), Object-Oriented Programming. Proceedings, 1994. XI, 535 pages. 1994.

Vol. 822: F. Pfenning (Ed.), Logic Programming and Automated Reasoning. Proceedings, 1994. X, 345 pages. 1994. (Subseries LNAI).

Vol. 823: R. A. Elmasri, V. Kouramajian, B. Thalheim (Eds.), Entity-Relationship Approach — ER '93. Proceedings, 1993. X, 531 pages. 1994.

Vol. 824: E. M. Schmidt, S. Skyum (Eds.), Algorithm Theory – SWAT '94. Proceedings. IX, 383 pages. 1994.

Vol. 825: J. L. Mundy, A. Zisserman, D. Forsyth (Eds.), Applications of Invariance in Computer Vision. Proceedings, 1993. IX, 510 pages. 1994.

Vol. 826: D. S. Bowers (Ed.), Directions in Databases. Proceedings, 1994. X, 234 pages. 1994.

Vol. 827: D. M. Gabbay, H. J. Ohlbach (Eds.), Temporal Logic. Proceedings, 1994. XI, 546 pages. 1994. (Subseries LNAI).

Vol. 828: L. C. Paulson, Isabelle. XVII, 321 pages. 1994.

Vol. 829: A. Chmora, S. B. Wicker (Eds.), Error Control, Cryptology, and Speech Compression. Proceedings, 1993. VIII, 121 pages. 1994.

Vol. 830: C. Castelfranchi, E. Werner (Eds.), Artificial Social Systems. Proceedings, 1992. XVIII, 337 pages. 1994. (Subseries LNAI).

Vol. 831: V. Bouchitté, M. Morvan (Eds.), Orders, Algorithms, and Applications. Proceedings, 1994. IX, 204 pages. 1994.

Vol. 832: E. Börger, Y. Gurevich, K. Meinke (Eds.), Computer Science Logic. Proceedings, 1993. VIII, 336 pages. 1994.

Vol. 833: D. Driankov, P. W. Eklund, A. Ralescu (Eds.), Fuzzy Logic and Fuzzy Control. Proceedings, 1991. XII, 157 pages. 1994. (Subseries LNAI).

Vol. 834: D.-Z. Du, X.-S. Zhang (Eds.), Algorithms and Computation. Proceedings, 1994. XIII, 687 pages. 1994.

Vol. 835: W. M. Tepfenhart, J. P. Dick, J. F. Sowa (Eds.), Conceptual Structures: Current Practices. Proceedings, 1994. VIII, 331 pages. 1994. (Subseries LNAI).

Vol. 836: B. Jonsson, J. Parrow (Eds.), CONCUR '94: Concurrency Theory. Proceedings, 1994. IX, 529 pages. 1994.

Vol. 837: S. Wess, K.-D. Althoff, M. M. Richter (Eds.), Topics in Case-Based Reasoning. Proceedings, 1993. IX, 471 pages. 1994. (Subseries LNAI).

Vol. 838: C. MacNish, D. Pearce, L. Moniz Pereira (Eds.), Logics in AI. Proceedings, 1994. IX, 413 pages. 1994. (Subseries LNAI).

Vol. 839: Y. G. Desmedt (Ed.), Advances in Cryptology - CRYPTO '94. Proceedings, 1994. XII, 439 pages. 1994.

Vol. 840: G. Reinelt, The Traveling Salesman. VIII, 223 pages. 1994.

Vol. 841: I. Prívara, B. Rovan, P. Ružička (Eds.), Mathematical Foundations of Computer Science 1994. Proceedings, 1994. X, 628 pages. 1994.

Vol. 842: T. Kloks, Treewidth. IX, 209 pages. 1994.

Vol. 843: A. Szepietowski, Turing Machines with Sublogarithmic Space. VIII, 115 pages. 1994.